# Algebra 1

Test Booklet

1·888·854·MATH (6284) - MathUSee.com
Sales@MathUSee.com

Sales@MathUSee.com – 1-888-854-MATH (6284) – MathUSee.com
Copyright © 2009 by Math-U-See

All rights reserved. No part of this book may be reproduced, stored in a retrieval system, or transmitted in any form by any means—electronic, mechanical, photocopying, recording or otherwise.

In other words, thou shalt not steal.

Printed in the United States of America

1115-0513

# Algebra 1 Tests

Tests are primarily for evaluating a student's progress. If a student does well on the test, then he is ready to move on to the next lesson. If he does not do well, spend more time on that lesson and master it before moving to the new material. (The solutions for the tests are in the instruction manual.) Math is sequential and builds from concept to concept and from lesson to lesson. Master the material in each lesson before moving to the next topic.

The simplest way to solve a multiple-choice problem is to pretend there are no answers given and simply solve it, and then find your answer among those offered. But it is also a good idea to estimate the answer before doing any calculations and eliminate several of the possibilities.

When the potential answers have been narrowed, then solve the problem and choose the correct answer.

You will find that this form of testing measures your reasoning abilities as well as your math knowledge. It also requires more than a cursory knowledge of the material being tested. Let me encourage you to look upon these tests, not merely as exams to conquer and pass, but also as an opportunity to learn and stretch your knowledge about this particular topic.

This booklet also contains three unit tests and a final exam, which are not written in multiple-choice format. They are designed to help you, as you move through the course, to remember what you have already learned.

## STANDARDIZED TESTS

The tests in this book have been written in the format most likely to be encountered on standardized tests. If the student is not familiar with the thinking skills employed in attacking multiple-choice answers, please spend some time explaining the different format and the different way of thinking that is required. Before attempting the SAT or ACT tests, it is recommended that the student complete *Algebra I*, *Geometry*, and most of *Algebra 2*.

—*Steve Demme*

# LESSON TEST 3

Circle your answer.

1. Solve for X:
   $-3X + 2 + 5X - 3 = 8 + 9$

   A. −9
   B. 9
   C. −2 1/4
   D. 4
   E. 2 1/4

2. Solve for D:
   $3D - 3 + 8 + D - D = 9 + 9 - 1$

   A. 4
   B. −3
   C. 11
   D. −4
   E. 3

3. Solve for B:
   $-6 + 2 + 3B + 4 = 2(4 + 1) - 1$

   A. 1
   B. −5
   C. 3
   D. −1
   E. −3

4. If X = 5, what is the value of $-2X + 2 + 5X + 8$?

   A. 55
   B. 25
   C. 41
   D. 21
   E. −45

5. If B = 3, what is the value of $(B+7) \times (B^2 - 10)$?

   A. −6
   B. 4
   C. 5
   D. −10
   E. 2

6. Solve for Q:
   $5Q - 9 - 6 = -1 \times 25$

   A. 2
   B. −2
   C. −4 2/5
   D. 8
   E. 4 2/5

7. Solve for Y:
   $-3 + Y + Y - 6 + 2 = 6 + 7$

   A. 12
   B. 6
   C. −6
   D. 21
   E. 10

8. Which equation has the largest value of X?

   A. $X - 3 = 9$
   B. $X + 3 = 9$
   C. $3X = 9$
   D. $X + 1 = 9$
   E. $X - 1 = 12$

LESSON TEST 3

9. Which equation has the smallest value of R?
   - A. R + 2R = 15
   - B. 2R + 3 = 15
   - C. R + 2R = 18
   - D. R + 5R = 15
   - E. R + 5R = 6

10. Which equations have the smallest value of Q?
    - I. 3Q − 4 = 20
    - II. 4Q − 3 = 17
    - III. 4Q + 3 = 23
    - IV. 4Q − 3Q = 21

    - A. II and III
    - B. III and IV
    - C. III only
    - D. IV only
    - E. I only

11. Solve for P:
    5 + P − 3 = 3(6) + 5P

    - A. −4
    - B. −2 2/3
    - C. 4
    - D. 5
    - E. −5

12. Solve for X:
    $$\frac{3}{4} + \frac{1}{2} = \frac{2}{3}X$$
    - A. 7
    - B. 2
    - C. 1
    - D. 1 7/8
    - E. 3

13. Solve for Y:
    $$\frac{3}{5}Y - \frac{1}{3} = \frac{1}{5}$$
    - A. 5/24
    - B. 8/9
    - C. 9/8
    - D. 1
    - E. 8

14. Solve for X:
    .09X − 1.8 = 2.25
    - A. 405
    - B. .45
    - C. 5
    - D. 45
    - E. 4.5

15. Solve for A:
    .6A + 15 = 7.2
    - A. 13
    - B. −13
    - C. 37
    - D. −1.3
    - E. 3.7

# LESSON TEST 4

Circle your answer.

1. What number is the GCF of 14, 16, and 28?
   A. 7
   B. 2
   C. 4
   D. 8
   E. 28

2. What number is the GCF of 14, 16, and 24?
   A. 8
   B. 2
   C. 6
   D. 4
   E. 16

3. What number is the GCF of 24, 36, and 40?
   A. 4
   B. 6
   C. 12
   D. 8
   E. 9

4. What number is the GCF of 26, 52, and 65?
   A. 3
   B. 2
   C. 4
   D. 15
   E. 13

5. Distribute 3(A + B + 6).
   A. 3AB + 6
   B. 18AB
   C. 3A + B + 6
   D. 3A + 3B + 6
   E. 3A + 3B + 18

6. Distribute 6(X − 2Y + 3 + Z).
   A. 6X − 4Y + 3 + 3Z
   B. 6X − 12Y + 18 + 6Z
   C. 4XY + 3Z
   D. 36XYZ
   E. −36XYZ

7. Distribute 2(3T − 5 + 4T + 3).
   A. 14T + 4
   B. 4Y − 16
   C. 6T + 4
   D. 14T − 4
   E. 10T

8. Distribute A(B + 4Q + 1).
   A. AB + 4AQ + A
   B. AB + 4AQ + 1
   C. AB + 5Q
   D. AB + 5AQ
   E. 5ABQ

LESSON TEST 4

9. What is the GCF of the terms in 10B − 15B = 40?
   A. 10
   B. 5
   C. 15
   D. 40
   E. 20

10. What is the GCF of the terms in 36X + 12Y = 24Z?
    A. 6
    B. 4
    C. 36
    D. 24
    E. 12

11. What is the GCF of the terms in 60A + 30D = 90?
    A. 10
    B. 90
    C. 30
    D. 60
    E. 20

12. Factor each term by the GCF: 18A + 24B = 30.
    A. 2(9A + 12B) = 30
    B. 2(9A + 12B) = 2(15)
    C. 6(3A + 4B) = 6(5)
    D. 24(A + B) = 30
    E. 6A + 8B = 10

13. Factor each term by the GCF: 15P − 25R = 35T.
    A. 3(5P − 5R) = 3(10)T
    B. 5(3P − 5R) = 5(7)T
    C. 5P − 5R = 5T
    D. 15(P − 10R) = 20T
    E. 15(P − 10R) = 15(2T)

14. Simplify 4G + 16H − 8J = 32.
    A. 2G + 8H − 4J = 16
    B. 12(G + H − J) = 32
    C. G + 4H − 2J = 32
    D. 4G + 2H − J = 4
    E. G + 4H − 2J = 8

15. Simplify 9X + 27Y = 3Z.
    A. X + 3Y = 3Z
    B. 3X + 9Y = Z
    C. X + 3Y = Z
    D. 36XY = 3Z
    E. 12XY = Z

# LESSON TEST 5

Use the graph for #1–5. Circle your answer.

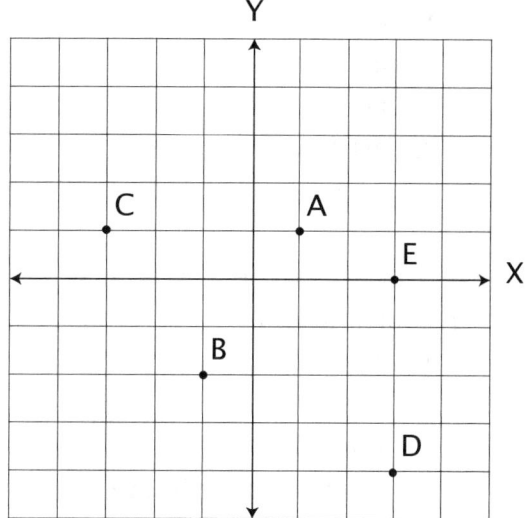

1. The coordinates of point A are:
   A. (1, –1)
   B. (–1, 1)
   C. (–1, –1)
   D. (1, 0)
   E. (1, 1)

2. The coordinates of point B are:
   A. (–2, –1)
   B. (–1, –2)
   C. (1, 2)
   D. (2, 1)
   E. (–1, 2)

3. The coordinates of point C are:
   A. (–1, 3)
   B. (–1, –3)
   C. (–3, 1)
   D. (1, 3)
   E. (3, 1)

4. The coordinates of point D are:
   A. (3, –4)
   B. (–4, 3)
   C. (–4, –3)
   D. (–3, –4)
   E. (–3, 4)

5. The coordinates of point E are:
   A. (–3, 0)
   B. (0, 3)
   C. (3, 0)
   D. (0, –3)
   E. (3, 1)

6. The point (6, –5) would lie in or at:
   A. quadrant I
   B. quadrant II
   C. quadrant III
   D. quadrant IV
   E. the origin

7. The point (15, 3) would lie in or at:
   A. quadrant I
   B. quadrant II
   C. quadrant III
   D. quadrant IV
   E. the origin

8. The point (0, 0) would lie in or at:
   A. quadrant I
   B. quadrant II
   C. quadrant III
   D. quadrant IV
   E. the origin

LESSON TEST 5

9. The word "Cartesian" is related to the word:
   A. cart
   B. artesian
   C. Descartes
   D. artificial
   E. carton

10. Every point on the X-axis has a Y-coordinate of:
    A. 1
    B. 2
    C. the same number as the X-coordinate
    D. 0
    E. none of the above

11. If the X-coordinate is –3 and the Y-coordinate is –8, where does the point lie?
    A. quadrant I
    B. quadrant II
    C. quadrant III
    D. quadrant IV
    E. the origin

12. Every point on the Y-axis has a X-coordinate of:
    A. 1
    B. 2
    C. the same number as the Y-coordinate
    D. 0
    E. none of the above

13. Analytic geometry combines:
    A. arithmetic and geometry
    B. logic and geometry
    C. algebra and geometry
    D. arithmetic and geography
    E. statistics and geometry

14. What is true about the following points?
    (3, –2), (0, –2), (–1, –2)
    A. They form a straight line.
    B. They are all in the 4th quadrant.
    C. They intersect the origin.
    D. They lie along the X-axis.
    E. They are all in the 3rd quadrant.

15. What is true about the following points?
    (4, 0), (4, 3), (4, –1), (5, 0)
    A. They form a straight line.
    B. They are all in either the 1st or 2nd quadrant.
    C. They lie along the X-axis.
    D. They cannot be connected with a straight line.
    E. They are all to the left of the Y-axis.

# LESSON TEST 6

Circle your answer.

1. Sue had finished three Christmas gifts before the first of December. After that she finished one gift each day.

   Which equation describes her work if G = gifts and D = days?
   - A. G = D + 3
   - B. D = G + 3
   - C. G = 3
   - D. G = 3D + 3
   - E. G = D + 4

2. Katie could already play five songs on her harp. She decided to learn two new songs each week.

   If S = songs and W = weeks, which equation expresses her goals?
   - A. W = 5S + 2
   - B. S = 5W + 2
   - C. S = 2W + 5
   - D. W = 2S + 5
   - E. S = W + 7

3. A team won two games in preseason play. Then the team won two more each week.

   How can the information be shown if G = games won and W = weeks?
   - A. W = 2G + 2
   - B. G = 2W + 2
   - C. G = 4W
   - D. 2G = 4W
   - E. G + 2 = 2

4. A scientist ran five computations to be sure his computer was programmed correctly. Then he ran two computations a day for six days.

   How many computations did he do in all?
   - A. 16
   - B. 13
   - C. 17
   - D. 32
   - E. 12

5. Bob did chores for his neighbor. The first day he earned $8. Then he earned $10 a day for two weeks.

   If he took one day off each week, how much did he earn?
   - A. $28
   - B. $80
   - C. $120
   - D. $148
   - E. $128

6. For Y = 4X − 1, substitute 3 for X and find the value of Y.
   - A. 6
   - B. 13
   - C. 3
   - D. −12
   - E. 11

ALGEBRA 1 LESSON TEST 6

LESSON TEST 6

7. A = 6B + 4. Substitute 0 for B and find the value of A.
   A. 0
   B. 4
   C. 10
   D. -4
   E. 6

8. R = T - 5. Substitute 2 for T and find the value of R.
   A. 6
   B. 13
   C. 3
   D. -12
   E. -3

9. Which of the following equations is true for (4, 13)? Substitute these values of X and Y into each one to see if the equation is true.
   A. Y = 3X + 1
   B. Y = 3X - 1
   C. X = 3Y + 1
   D. Y = X - 4
   E. X = 3Y - 2

10. Which of the following is true for (-2, -6)?
    A. Y = X + 4
    B. Y = 3X - 2
    C. X = Y - 4
    D. Y = X - 4
    E. X = 3Y - 2

The following five questions may be counted as extra credit if the teacher wishes. Use the graph for #11–15.

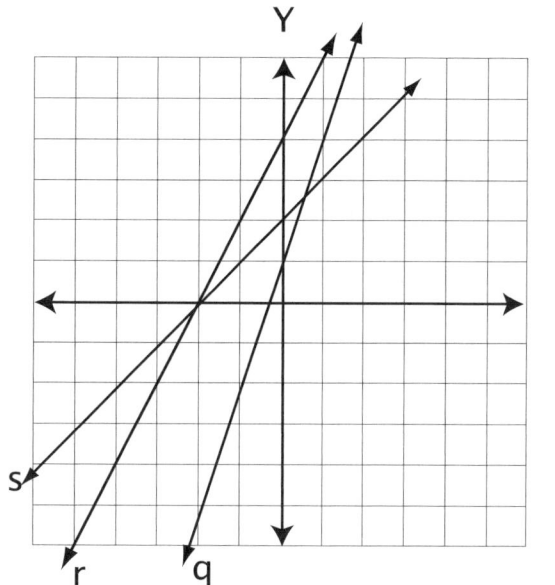

11. Points (0, 1) and (–1, –2) are on line *q*. Substitute the values of X and Y for these points into the following equations to find which one represents line *q*.
    A. Y = 4X
    B. Y = 3X + 1
    C. Y = X + 6
    D. Y = 6X + 1
    E. Y = 5X

12. Which line includes points that satisfy the equation X = 0?
    A. *q*
    B. *r*
    C. *s*
    D. X-axis
    E. Y-axis

13. Substitute points from line *s* to determine which of the following is the equation for line *s*.
    A. Y = X + 3
    B. Y = 2X
    C. Y = X + 2
    D. Y = X – 3
    E. Y = 3X + 1

14. Which line includes points that satisfy the equation Y = 2X + 4? (Hint: see #11.)
    A. *q*
    B. *r*
    C. *s*
    D. X-axis
    E. Y-axis

15. What is the equation for the X-axis?
    A. X = 0
    B. Y = X
    C. Y = X + 1
    D. X = 1
    E. Y = 0

LESSON TEST 6

# LESSON TEST 7

Circle your answer.

1. A line with a positive slope:
   A. is always a vertical line
   B. is always a horizontal line
   C. slants up to the right
   D. slants down to the right
   E. none of the above

2. A line with a negative slope:
   A. is always a vertical line
   B. is always a horizontal line
   C. slants up to the right
   D. slants down to the right
   E. none of the above

3. Parallel lines have the same:
   A. slope
   B. Y-intercept
   C. X-intercept
   D. formula
   E. none of the above

4. In the formula Y = 3X + 4, the slope is:
   A. X
   B. Y
   C. 3
   D. 4
   E. none of the above

5. In the formula Y = 3X + 4, the Y-intercept is:
   A. X
   B. Y
   C. 3
   D. 4
   E. 0

6. If a line has a rise of 3 and a run of 4, what is its slope?
   A. 3
   B. 4
   C. 3/4
   D. 4/3
   E. 7

7. If a line has a rise of 2 and a run of 2, what is its slope?
   A. 1
   B. 2
   C. 4
   D. 0
   E. −2

8. If a line goes through the point (0, 4), what is its Y-intercept?
   A. 1
   B. −4
   C. 0
   D. 4
   E. cannot be determined from the information given

ALGEBRA 1 LESSON TEST 7

LESSON TEST 7

Use these diagrams for #9–15.

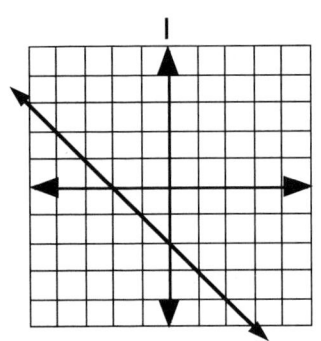 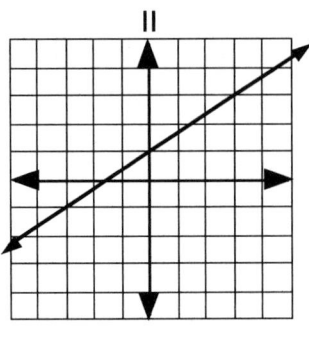 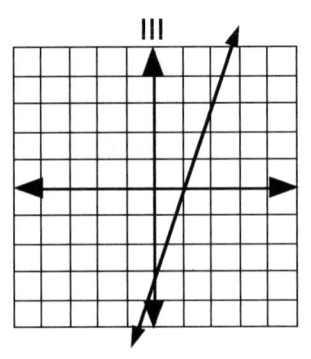 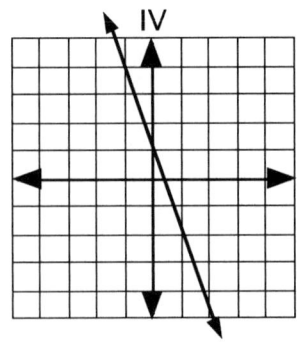

9. Which line shown appears to have a slope of 2/3?
   A. I
   B. II
   C. III
   D. IV
   E. none of the above

10. Which line shown appears to have a slope of –1?
    A. I
    B. II
    C. III
    D. IV
    E. none of the above

11. Which line shown appears to have a slope of 2?
    A. I
    B. II
    C. III
    D. IV
    E. none of the above

12. Which line shown appears to have a slope of 3?
    A. I
    B. II
    C. III
    D. IV
    E. none of the above

13. Which line shown appears to have a Y-intercept of 3?
    A. I
    B. II
    C. III
    D. IV
    E. none of the above

14. Which line shown appears to have a Y-intercept of –2?
    A. I
    B. II
    C. III
    D. IV
    E. none of the above

15. Which line shown appears to have a Y-intercept of –3?
    A. I
    B. II
    C. III
    D. IV
    E. none of the above

# LESSON TEST 8

Circle your answer.

1. Sally gained three pounds over vacation, and then she lost one pound per week.

    If P = pounds and W = weeks, which is the correct equation?
    - A. P = (-1)W + 3
    - B. P = W - 3
    - C. P = -3W + 1
    - D. P = -10W + 50
    - E. P = W + 3

2. The politician's popularity rating was 50%. Then he began to lose popularity at a rate of 10% per week.

    If R = current rating and W = weeks, which is the correct equation?
    - A. R = 10W - 50B
    - B. R = -50W + 10
    - C. R = 50W - 10
    - D. R = -10W + 50
    - E. R = -10W - 50

3. George started his business with $4. Unfortunately, he lost $5 a day.

    If M = money and D = days, which equation shows his financial condition?
    - A. D = -5M + 4
    - B. D = 5M - 4
    - C. M = 5D + 4
    - D. M = -5D - 4
    - E. M = -5D + 4

4. Jim was already in debt $5 when he went into business. However, he was able to make $4 a day.

    If M = money and D = days, which is the correct equation?
    - A. M = -4D + 5
    - B. M = 4D + 5
    - C. M = 4D - 5
    - D. D = 4M - 5
    - E. D = -4M + 5

5. Shelly tended to lose things. She lost four things last month. This month she lost her belongings at the rate of three per week.

    If T = things and W = weeks, choose the correct equation.
    - A. W = -3T - 4
    - B. T = -3 - 4W
    - C. T = -3W + 4
    - D. T = -3W - 4
    - E. T = -4W - 3

6. What is the slope of the line described by Y = 4X + 1?
    - A. 4
    - B. 1
    - C. 5
    - D. 3
    - E. -4

LESSON TEST 8

7. What is the slope of the line described by Y = −2X + 6?
   A. 6
   B. 2
   C. −2
   D. −6
   E. 4

8. What is the slope of the line described by Y = 1/2 X − 5?
   A. −5
   B. 15
   C. 2
   D. −1/2
   E. 1/2

9. What is the Y-intercept of a line that includes the following points: (0, 2), (1, 5), (−1, −1)?
   A. 0
   B. 2
   C. 1
   D. 5
   E. 8

10. Sketch the line described in #9. What is its slope?
    A. 3
    B. 1
    C. 1/3
    D. 2
    E. 5

11. What is the equation for the line described in #9 and #10?
    A. Y = 2X + 3
    B. Y = 5X
    C. Y = 3X + 2
    D. Y = 1/3 X + 2
    E. Y = 2X + 1/3

Use the graph for #12–15.

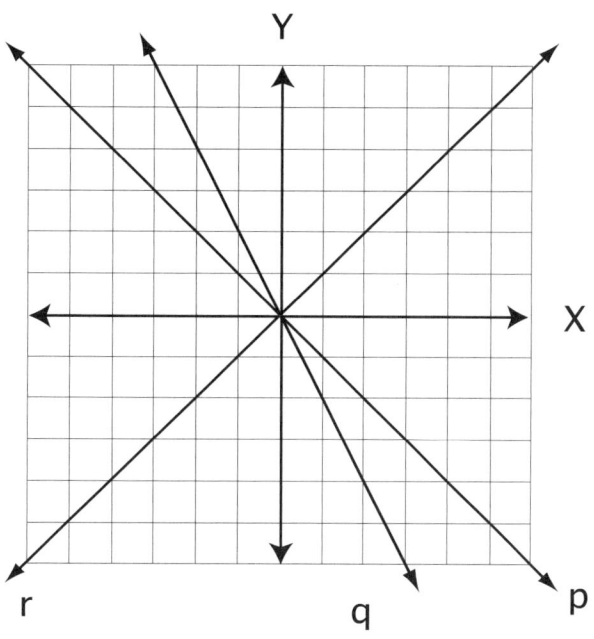

12. The slope of line *r* is:
    A. −1
    B. 1
    C. 2
    D. 1/2
    E. −2

13. The slope of line *q* is:
    A. 4
    B. −2
    C. 2
    D. 1/2
    E. −1/2

14. The slope of the X-axis is:
    A. 0
    B. 1
    C. −1
    D. 1/2
    E. .01

15. The slope of line *p* is:
    A. −1
    B. 1
    C. 2
    D. 1/2
    E. −2

# LESSON TEST 9

Circle your answer.

1. Parallel lines have the same:
   A. slope
   B. intercept
   C. endpoint
   D. length
   E. equation

2. Line $q$ and line $r$ each have a slope of 4. They are:
   A. curved
   B. the same
   C. parallel
   D. in the same quadrant
   E. none of the above

Use the graph to sketch the problems as needed.

3. A line with points (3, 3) and (4, 5) has a slope of:
   A. 5
   B. 3
   C. 4
   D. 2
   E. 1

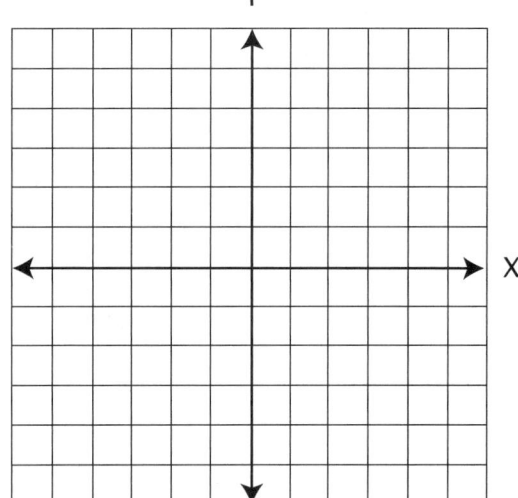

4. A line with points (−4, 0) and (−3, 1) has a slope of:
   A. −1
   B. 3/4
   C. 1
   D. 2
   E. −3

6. Which of the following is parallel to $2Y = 6X + 4$?
   A. $Y = -3X + 3$
   B. $Y = 3X + 4$
   C. $Y = X + 3$
   D. $Y = -2X - 3$
   E. $Y = 1/2\,X + 3$

5. A line with points (−3, 1) and (0, 0) has a slope of:
   A. 1/3
   B. −1/3
   C. 3
   D. −3
   E. 1

7. Which of the following is parallel to $3Y = 6X + 3$?
   A. $Y = 2X + 10$
   B. $Y = 6X + 4$
   C. $Y = 6X + 2$
   D. $Y = 1/2\,X + 8$
   E. $Y = -3X + 2$

LESSON TEST 9

8. Which of the following is in the standard form of the equation of a line?
   A. $3X + 2Y = 3$
   B. $3X + 2Y - 3 = 0$
   C. $X = Y$
   D. $Y = 2X + 3$
   E. $X = 2Y + 4$

9. At what point does the line $Y = 2X + 6$ intercept the Y-axis?
   A. 2
   B. 8
   C. -2
   D. 1/6
   E. 6

10. What is $Y = 2X + 6$ written using the standard form of the equation of a line?
    A. $Y + 2X = 6$
    B. $Y - 2X - 6 = 0$
    C. $X = 2Y + 6$
    D. $Y = X + 6$
    E. $2X - Y = -6$

11. At what point does the line $Y = -1/2\ X + 2$ intercept the Y-axis?
    A. -1
    B. -1/2
    C. 1
    D. 2
    E. -2

Use the graph for #12–14.

12. An equation for line *h* is:
    A. $Y = -2X - 4$
    B. $Y = -3X + 2$
    C. $Y = 2 + 3X$
    D. $Y = 2X - 4$
    E. $Y = 2X + 3$

13. The equation $Y = -X$ refers to line:
    A. f
    B. g
    C. h
    D. X
    E. Y

14. The slope of line *g* is:
    A. -2
    B. 3
    C. 2/3
    D. -2/3
    E. -3

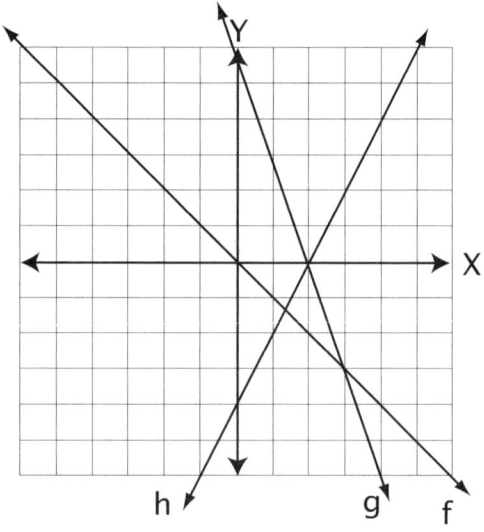

15. The Y-intercept of the X-axis is:
    A. X
    B. Y
    C. 0
    D. 1
    E. infinity

LESSON TEST

# 10

Circle your answer.

1. Line *q* has a slope of 4 and line *r* has a slope of −1/4.
   They are:
   A. parallel
   B. the same
   C. perpendicular
   D. supplementary
   E. none of the above

2. The slope of a line perpendicular to a given line is:
   A. the same as the given slope
   B. the negative of the given slope
   C. twice the given slope
   D. the inverse of the given slope
   E. none of the above

Use the graph to sketch the problems as needed.

3. A line with points (−2, 2) and (2, 4) has a slope of:
   A. 2
   B. 1/2
   C. 1
   D. −2
   E. −1/2

4. What is the Y-intercept of the line in #3?
   A. −6
   B. 2
   C. 1/2
   D. 3
   E. −3

5. What is the equation for #3 in slope-intercept form?
   A. X − 2Y = 6
   B. Y = 3X + 1/2
   C. Y = 2X + 3
   D. Y = 1/2 X + 3
   E. 2Y − 6X = 1

6. What is the standard form of the equation of a line for #3?
   A. −X + 2Y = 6
   B. Y = 3X + 1/2
   C. Y = 2X + 3
   D. Y = 1/2 X + 3
   E. 2Y − 6X = 1

ALGEBRA 1   LESSON TEST 10                                                27

LESSON TEST 10

7. What is the slope of a line perpendicular to the line described in #3?
   A. 1/2
   B. 2
   C. -2
   D. -1/2
   E. -1/3

8. A line with points (-1, 3) and (1, -1) has a slope of:
   A. 2
   B. 3
   C. -2
   D. 1/2
   E. -1/2

9. What is the Y-intercept of the line in #8?
   A. (1, 0)
   B. (0, -1)
   C. (-1, 0)
   D. (1, 1)
   E. (0, 1)

10. Write the equation for #8 in slope-intercept form.
    A. Y = -2X + 1
    B. Y + 2X = 1
    C. Y = 2X + 1
    D. Y - 2X = 1
    E. X = -2Y + 1

11. Write the answer for #10 in the standard form of the equation of a line.
    A. Y = -2X + 1
    B. 2X + Y = 1
    C. Y = 2X + 1
    D. Y - 2X = 1
    E. X = -2Y + 1

12. What is the slope of a line perpendicular to the line described in #8?
    A. -1/2
    B. 2
    C. -2
    D. 1/2
    E. -1

13. Which of the following is perpendicular to Y = 1/3 X + 6?
    A. Y = -1/3 X + 6
    B. Y = -1/3 X - 6
    C. Y = 3X - 6
    D. Y = 3X + 6
    E. Y = -3X + 6

14. Which of the following is perpendicular to Y = -4X - 1?
    A. Y = -2X - 1
    B. Y = -4X + 1
    C. Y = 4X - 1
    D. Y = -1/4 X - 1
    E. Y = 1/4 X - 1

15. What is the slope of a line perpendicular to 3Y + 6X = 12?
    A. -2
    B. 2
    C. 1/2
    D. -1/2
    E. 1/6

# LESSON TEST 11

Circle your answer.

1. Two points on line *p* have coordinates (2, 1) and (5, 3). The slope of the line is:
    A. 2
    B. 3/2
    C. 1
    D. 2/3
    E. 4

2. The line passing through points (4, 0) and (–2, 1) has a slope of:
    A. –6
    B. –1/6
    C. 1/2
    D. 2
    E. 1/6

3. The two points and (–3, –2) and (4, 8) are part of a line with a slope of:
    A. 7/10
    B. 1/6
    C. 6
    D. –6
    E. 10/7

4. How would 2X + Y = 13 be written in slope-intercept form?
    A. Y = 2X + 13
    B. 2X + Y – 13 = 0
    C. Y = 2X – 13
    D. X = 2Y + 13
    E. Y = –2X + 13

5. How would –3X + 4Y = 8 be written in slope-intercept form?
    A. Y = 3/4 X + 8
    B. Y = 3/4 X + 2
    C. Y = –3/4 X + 2
    D. –3X + 4Y – 8 = 0
    E. X = –4/3 Y + 8/3

6. How would 2X – 2Y – 6 = 0 be written in slope-intercept form?
    A. 2X – 2Y = 6
    B. –2Y = 2X = 6
    C. Y = X – 3
    D. Y = –2X – 3
    E. Y = X + 3

7. What is the slope of the line described by Y = 6X + 2?
    A. 6
    B. 2
    C. 3
    D. –6
    E. 12

8. What is the slope of the line described by 2X + 3Y = 4?
    A. 2/3
    B. –2/3
    C. 3/2
    D. 2
    E. 3

ALGEBRA 1 LESSON TEST 11

LESSON TEST 11

9. What is the slope of the line described by $-4X + 2Y = 16$?
   A. $-2$
   B. $-4$
   C. $4$
   D. $2$
   E. $16$

10. Which of the following give enough information to write the slope-intercept equation for a line?
    I. one point
    II. a point and the slope
    III. two points
    IV. the slope

    A. II only
    B. I or II
    C. II or III
    D. III only
    E. IV only

11. Given a slope of 3 and the point (2, 1), find the Y-intercept of the line.
    A. 5
    B. $-5$
    C. 1
    D. $-1$
    E. 3

12. Given a slope of $-1$ and the point $(-2, -2)$, find the Y-intercept of the line.
    A. $-4$
    B. 4
    C. $-5$
    D. 0
    E. 3

13. The equation for a line passing through (6, 3) and (4, 1) could be written in slope-intercept form as:
    A. $Y = X + 3$
    B. $Y = X + 2$
    C. $Y = 2X - 3$
    D. $Y = 2X + 3$
    E. $Y = X - 3$

14. The equation for a line passing through $(-4, 6)$ and $(1, 0)$ could be written in slope-intercept form as:
    A. $Y = \frac{6}{5}X + \frac{6}{5}$
    B. $Y = -\frac{6}{5}X + \frac{6}{5}$
    C. $Y = -\frac{6}{5}X - \frac{6}{5}$
    D. $Y = -\frac{5}{6}X + \frac{6}{5}$
    E. $Y = 6X + 6$

15. Which equation describes a line passing through (2, 3) and (4, 1) and having a slope of $-1$?
    A. $Y = -X + 5$
    B. $Y = X - 5$
    C. $Y = -X - 5$
    D. $Y = -\frac{1}{2}X + 5$
    E. $Y = -2X + 5$

UNIT TEST   **Lessons 1-11** (100 points possible)

I. Simplify. (4 points each)

1. $-3^2$

2. $-2 + 3^2 - 1 \times 4$

3. $|3 - 2| - |1 - 4|$

II. Solve. (6 points each)

1. $3X - 2 + 2X = 4 - X$

2. $\frac{1}{2}B + \frac{1}{3} = \frac{2}{9}$

3. $.03Y + 1 = 4.3$

III. Which point is found in the second quadrant? (5 points)

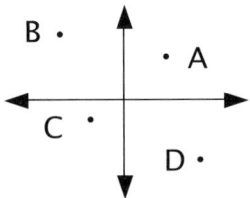

UNIT TEST I

IV. Label each of the following as an example of the commutative, associative, or distributive property. (4 points each)

1. (r + s) + t = r + (s + t)

2. B(C + D) = BC + BD

3. 3 + 8 = 8 + 3

V. Graph the following lines. (6 points each)

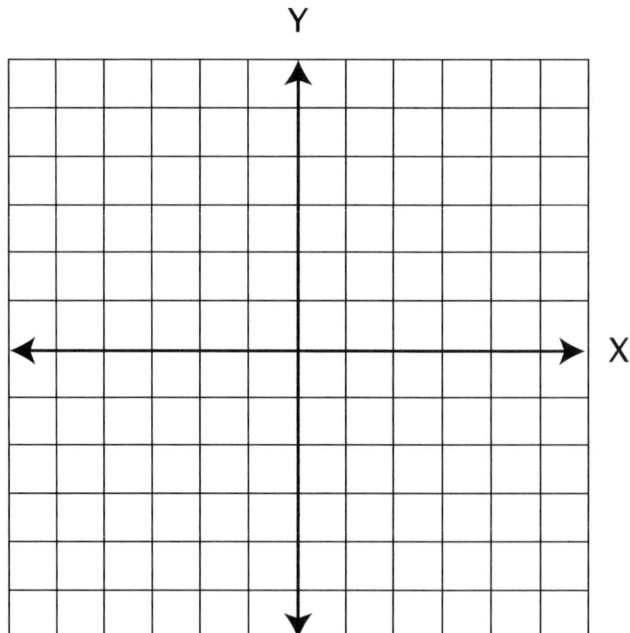

1. X = −1

2. Y = 2X − 1

VI. Give the slope and intercept of the following line and graph it: Y + 3 = −2X. (10 points)

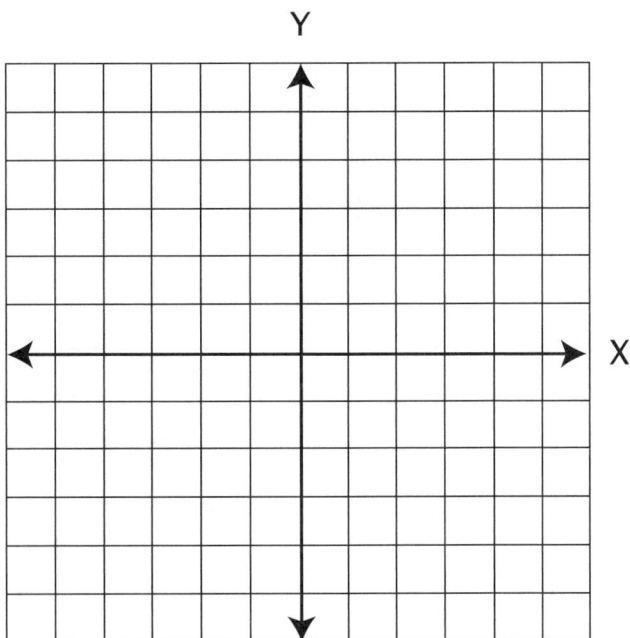

VII. Susan started her lemonade stand with a loan of $2. She was able to make $3 a day. If M = money and D = days, write an equation for the line that represents her financial condition. (5 points)

UNIT TEST I

VIII. Write the equation of the line through (2, 1) that is perpendicular to line Y = 3X. Graph both lines. (10 points)

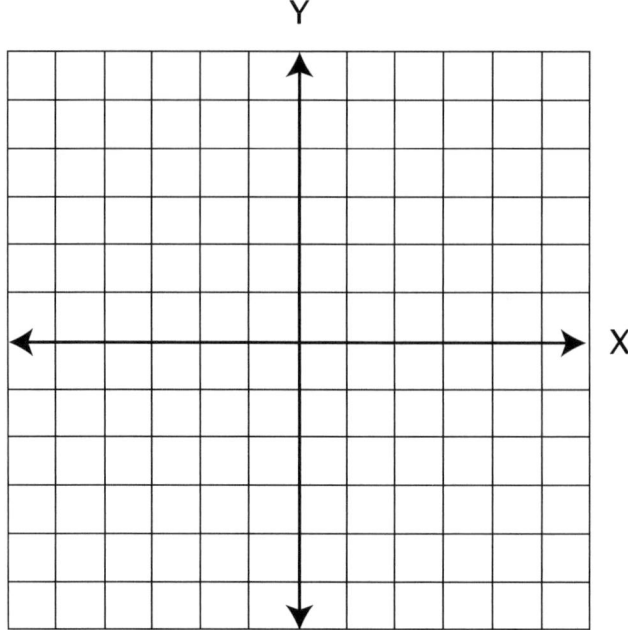

IX. Find the equation for a line passing through (2, 1) and (0, 4). (10 points)

X. Which of the following lines are parallel? (6 points)

a. Y = 3X + 7
b. Y = −71 + 3X
c. Y + 3X = 7
d. 3X = Y + 12

# LESSON TEST 12

Circle your answer.

1. Which of the following equations yields a dotted line when graphed?
   A. $Y \leq 3X + 1$
   B. $2Y > X + 4$
   C. $Y + 6X = 3$
   D. $Y \geq 3X + 4$
   E. $Y = X$

2. $-2Y > 4X + 8$ is the same as:
   A. $Y > -2X - 4$
   B. $-2Y > X + 4$
   C. $Y > 2X + 4$
   D. $Y < 2X + 4$
   E. $Y < -2X - 4$

3. $3Y < 6X - 6$ is the same as:
   A. $3Y > 6X - 6$
   B. $Y < 2X - 2$
   C. $Y > 2X - 2$
   D. $Y > 2X + 2$
   E. $Y \leq 2X - 2$

4. $Y \leq X - 4$ lies in all quadrants except:
   A. I and II
   B. III
   C. II
   D. I
   E. IV

5. The sign of an inequality should be reversed for which operations?
   I. adding a negative number
   II. multiplying by a negative number
   III. multiplying by a positive number
   IV. dividing by a positive number
   V. dividing by a negative number

   A. I, IV, V
   B. II, III, IV, V
   C. II, V
   D. IV, V
   E. I, II, III, IV, V

6. $2Y > 6X + 2$ is the same as:
   A. $Y < 3X + 1$
   B. $Y > 3X + 1$
   C. $Y > 3X + 2$
   D. $Y > 2X + 1$
   E. $X > 3Y + 1$

7. Which graph on the following page is the graph of the inequality in #6?
   A. I
   B. II
   C. III
   D. IV
   E. V

8. $-4Y + 8X > 16$ is the same as:
   A. $Y < -2X - 4$
   B. $Y > -2X + 4$
   C. $Y > 2X - 4$
   D. $Y < 2X + 4$
   E. $Y < 2X - 4$

LESSON TEST 12

I
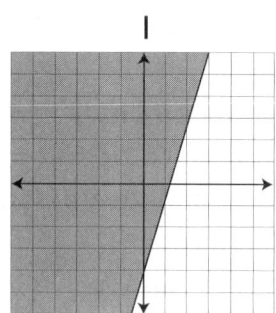

II
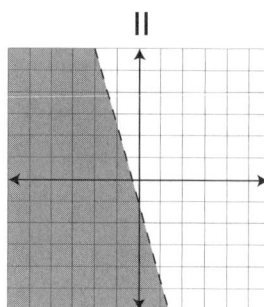

III
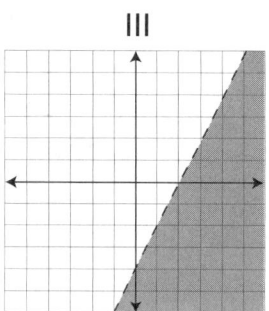

IV
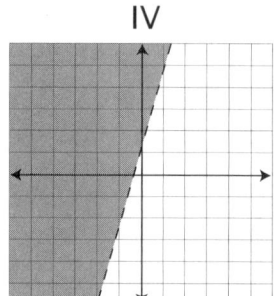

V
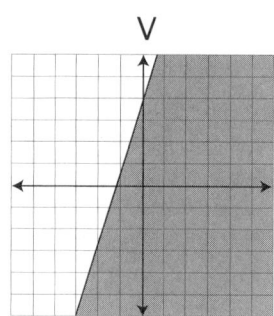

Use the graph for #10–15.

9. Which graph is the graph of the inequality in #8?
   A. I
   B. II
   C. III
   D. IV
   E. V

10. $-3Y + 9X \leq 12$ is the same as:
    A. $Y \geq 3X + 4$
    B. $Y \leq 3X + 4$
    C. $Y \leq -3X + 4$
    D. $Y \geq 3X - 4$
    E. $Y \geq -3X + 4$

11. Which graph above is the graph of the inequality in #10?
    A. I
    B. II
    C. III
    D. IV
    E. V

12. $3Y - 9X \leq 12$ is the same as:
    A. $Y \geq 3X + 4$
    B. $Y \leq 3X + 4$
    C. $Y \leq -3X + 4$
    D. $Y \geq 3X - 4$
    E. $Y \geq -3X + 4$

13. Which graph is the graph of the inequality in #12?
    A. I
    B. II
    C. III
    D. IV
    E. V

14. $-2Y > 6X + 2$ is the same as:
    A. $Y < -3X - 1$
    B. $Y > -3X + 1$
    C. $Y > 3X - 1$
    D. $Y > -3X - 1$
    E. $X < -3X + 1$

15. Which graph is the graph of the inequality in #14?
    A. I
    B. II
    C. III
    D. IV
    E. V

# LESSON TEST 13

Circle your answer.

1. How many points in a line satisfy the equation of that line?
   A. infinite number
   B. 1
   C. 2
   D. varies for different equations
   E. none

2. When two different lines intersect, how many points are common to both lines?
   A. infinite number
   B. 1
   C. 2
   D. varies for different equations
   E. none

3. To solve for two different equations, find the point(s) where:
   A. they cross the X-axis
   B. they cross the Y-axis
   C. X has the same value as Y
   D. the lines intersect
   E. either line intersects the origin

4. If two lines $a$ and $b$ intersect at point (3, 4), which of the following is not true?
   A. (3, 4) is a point on line $a$.
   B. (3, 4) is a point on line $b$.
   C. (3, 4) satisfies the equation for line $a$.
   D. (3, 4) satisfies the equation for line $b$.
   E. (3, 4) will not satisfy either equation.

Use the graph for #5–9.

5. Which line represents $Y = -4X - 2$?
   A. $a$
   B. $b$
   C. $c$
   D. $d$
   E. none

6. Which line represents $Y = X + 3$?
   A. $a$
   B. $b$
   C. $c$
   D. $d$
   E. none

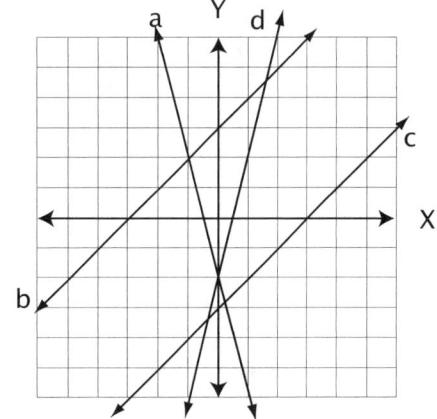

7. What point satisfies $Y = -4X - 2$ and $Y = X + 3$?
   A. (−2, −2)
   B. (0, −2)
   C. (−1, 2)
   D. (−2, 0)
   E. (2, −2)

ALGEBRA 1 LESSON TEST 13

LESSON TEST 13

Use the graph on previous page for #8–9.

8. Which line represents Y = 4X + 2?
   A. a
   B. b
   C. c
   D. d
   E. none

9. What point satisfies Y = –4X – 2 and Y = 4X – 2?
   A. (–2, –2)
   B. (0, –2)
   C. (–1, 2)
   D. (–2, 0)
   E. (2, –2)

Use this graph for #10–15.

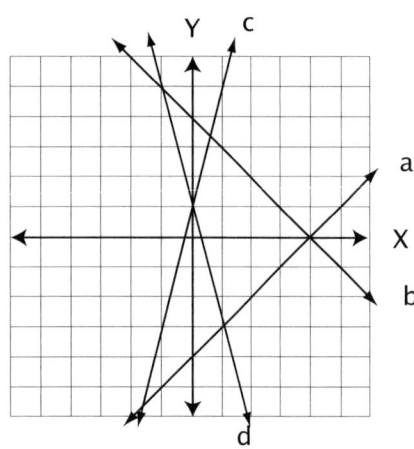

10. Which line represents Y = –X + 4?
    A. a
    B. b
    C. c
    D. d
    E. none

11. Which line represents Y = X – 4?
    A. a
    B. b
    C. c
    D. d
    E. none

12. Which line represents Y = 4X + 1?
    A. a
    B. b
    C. c
    D. d
    E. none

13. Which line represents Y = –4X + 1?
    A. a
    B. b
    C. c
    D. d
    E. none

14. What point satisfies Y = X – 4 and Y = –X + 4?
    A. (–2, 3)
    B. (0, 4)
    C. (0, 1)
    D. (1, 3)
    E. (4, 0)

15. What point satisfies Y = –4X + 1 and Y = 4X + 1?
    A. (1, 3)
    B. (0, 1)
    C. (4, 0)
    D. (1, 0)
    E. (2, 0)

# LESSON TEST 14

Circle your answer.

1. Simultaneous equations may be solved by:
   A. the distributive property
   B. the slope-intercept formula
   C. extrapolating
   D. substitution
   E. factoring

2. Given $Y - 2 = X$ and $Y + X = 8$, what could be substituted for X in the second equation?
   A. $Y + 2$
   B. $Y - 2$
   C. $Y - 8$
   D. $Y = 2$
   E. $2Y$

3. Which point satisfies $Y - 2 = X$ and $Y + X = 8$?
   A. (8, 0)
   B. (0, 8)
   C. (0, 0)
   D. (5, 3)
   E. (3, 5)

4. Given $Y + 5 = X$ and $8Y + X = 2$, what could be substituted for X in the first equation?
   A. $8Y + 2$
   B. $X - 2$
   C. $8Y - 2$
   D. $X + 4$
   E. $-8Y + 2$

5. Which point satisfies $Y + 5 = X$ and $8Y + X = 2$?
   A. (4 2/3, -1/3)
   B. (-1/3, 4 2/3)
   C. (2, -3)
   D. (-3, 2)
   E. (3/7, 5/7)

6. When solving simultaneous equations, first find the value of one variable, and then:
   A. multiply by that variable
   B. subtract the value from both sides
   C. use the reciprocal of the other variable
   D. substitute to find the other variable
   E. try to determine the slope of the line

7. A problem with solving simultaneous equations by graphing is:
   A. no problem—it is the preferred method
   B. that the results are incorrect if one number is zero
   C. that points cannot be found for both lines
   D. that it is very difficult and time consuming
   E. that the answer may be an estimate, especially for fractional values

ALGEBRA 1 LESSON TEST 14

LESSON TEST 14

8. Which point satisfies X + Y = 8 and 2X − Y = 7?
   A. (3, 5)
   B. (5, 3)
   C. (15, −7)
   D. (1, 7)
   E. (−5, 3)

9. Which point satisfies 2X + Y = 14 and Y = X + 2?
   A. (4, 6)
   B. (8, 10)
   C. (−8, −6)
   D. (6, 8)
   E. (4, −6)

10. Which point satisfies 2X + Y = 2 and Y − X = −1?
    A. (0, 1)
    B. (1, 0)
    C. (1/3, −1/3)
    D. (3, 2)
    E. (3, −2)

11. Which point satisfies Y = 6X + 4 and X = 2Y + 3?
    A. (−2, −1)
    B. (−1, −2)
    C. (−1/2, −1/4)
    D. (1/2, 1/4)
    E. (2, −1)

12. Stephen left at 6:00 AM and traveled 470 miles at a rate of 55 mph. About when will he arrive at his destination?
    A. 2:30 AM
    B. 8:00 AM
    C. 8:00 PM
    D. 4:00 PM
    E. 2:30 PM

13. If Stephen's vehicle used 23.5 gallons of gasoline, how many miles per gallon did he get? (See #12.)
    A. 45 mpg
    B. 2 mpg
    C. 8.5 mpg
    D. 20 mpg
    E. none of the above

14. What is the next number in this series? 4, 9, 16, 25, . . .
    A. 30
    B. 35
    C. 41
    D. 36
    E. 50

15. In 4X − 2Y = 12, which value of X yields the largest value of Y?
    A. 0
    B. −10
    C. 3
    D. 1
    E. 2

# LESSON TEST 15

Circle your answer.

1. Simultaneous equations may be solved by:
    I. graphing
    II. substitution
    III. elimination
    IV. extrapolating
    V. factoring

    A. I and V only
    B. V only
    C. III and IV only
    D. I, II and III
    E. I and II only

2. A good first step in solving $2X - 3Y = 4$ and $4X = Y + 5$ by elimination would be to:
    A. add the equations
    B. multiply both by 2
    C. make sure both are in the same form
    D. multiply the second by -3
    E. substitute 2X for Y in the second

3. Before solving $X + 3Y = 6$ and $2X + 4Y = 7$ by elimination, it would be helpful to multiply the first equation by:
    A. 4
    B. -2
    C. 7
    D. 6
    E. -4

4. Before solving $-X - 2Y = 3$ and $4X - 5Y = 4$ by elimination, it would be helpful to multiply the first equation by:
    A. 5/2
    B. 2/5
    C. 1/4
    D. 2
    E. 4

5. Given $X - 3Y = 6$ and $X + 3Y = 12$, what is the value of X?
    A. 3
    B. 6
    C. 9
    D. 18
    E. -3

6. What is the value of Y for #5?
    A. 3
    B. 1
    C. -1
    D. 0
    E. -3

7. If lines were drawn on a graph for the equations in #5, at what point would they intersect?
    A. (0, 1)
    B. (3, -1)
    C. (6, 0)
    D. (9, 1)
    E. cannot be determined from the information given

LESSON TEST 15

8. Given 2X + 4Y = 13 and 3X + Y = 2, what is the value of X?
   A. 2/3
   B. −2/3
   C. 3
   D. 1/2
   E. −1/2

9. What is the value of Y for #8?
   A. 3 1/2
   B. 3
   C. 2 11/12
   D. 1 3/4
   E. −1/2

10. If lines were drawn on a graph for the equations in #8, at what point would they intersect?
    A. (−1/2, 3 1/2)
    B. (−1/2, 3)
    C. (2/3, 2 11/12)
    D. (3, 1 3/4)
    E. cannot be determined

11. Which point satisfies X − 2Y = 4 and X + 6Y = 12?
    A. (1, 6)
    B. (2, 8)
    C. (6, 1)
    D. (8, 2)
    E. (−6, 1)

12. Which point satisfies −3X − Y = 9 and 2X + Y = 3?
    A. (6, −9)
    B. (−6, 9)
    C. (12, 9)
    D. (−12, 27)
    E. (−12, 9)

13. Use algebraic symbols to represent the following:

    Six times a number, minus four times the number, equals eight times the number, divided by four.

    A. 6N − 4 = N ÷ 4
    B. 6(N − 4N = 8N ÷ 4)
    C. 6(N − 4) = 8(N ÷ 2)
    D. 6N − 4 = 8N ÷ 4
    E. 6N − 4N = 8N ÷ 4

14. What is the next number in this series? 49, 64, 81, . . .
    A. 9
    B. 100
    C. 90
    D. 162
    E. 98

15. In 2X + 3Y = 6, which value of X yields the largest value of Y?
    A. 0
    B. 1
    C. 5
    D. −5
    E. 8

42  ALGEBRA 1

# LESSON TEST 16

Circle your answer.

1. P + N = 8, where P = pennies and N = nickels. If the value of the coins is 28 cents, what other equation is necessary to find the number of each coin.
   A. P + 5N = 28
   B. 5P + N = 28
   C. P + 28 = 5N
   D. (1 + 5)N = P
   E. 8 + N = 28

2. There are seven coins, all either nickels or dimes. The value of the coins is 50 cents. How many nickels are there?
   A. 3
   B. 4
   C. 7
   D. 10
   E. 5

3. How many dimes are there in the problem described in #2?
   A. 3
   B. 4
   C. 7
   D. 10
   E. 5

4. There are 13 coins, all either nickels or dimes. The value of the coins is $1.10. How many dimes are there?
   A. 4
   B. 5
   C. 10
   D. 9
   E. 8

5. How many nickels are there in the problem described in #4?
   A. 4
   B. 5
   C. 10
   D. 9
   E. 8

6. There are 13 coins, all either nickels or quarters. The value of the coins is $1.85. How many quarters are there?
   A. 10
   B. 13
   C. 5
   D. 6
   E. 7

7. How many nickels are there in the problem described in #6?
   A. 10
   B. 13
   C. 5
   D. 6
   E. 7

8. Coin problems are solved using:
   A. elimination
   B. factoring
   C. estimation
   D. squaring
   E. none of the above

ALGEBRA 1 LESSON TEST 16

LESSON TEST 16

9. There are 10 coins, all either dimes or quarters. The value of the coins is $2.05. How many dimes are there?
   A. 2
   B. 3
   C. 7
   D. 5
   E. 10

10. How many quarters are there in the problem described in #9?
    A. 2
    B. 3
    C. 7
    D. 5
    E. 10

11. The government minted two new coins. Coin A is worth $.30 and coin B is worth $.75. Jeff has five of the coins in his pocket with a total value of $2.40. How many of coin B does he have?
    A. 4
    B. 3
    C. 1
    D. 5
    E. 2

12. How many of coin A does Jeff have (#11)?
    A. 4
    B. 3
    C. 1
    D. 5
    E. 2

13. Lisa started with $50 and saved $15 a week. Which equation describes the growth of her savings with X as the number of weeks and Y as the total dollars?
    A. $X = 15Y + 50$
    B. $Y = 15X + 50$
    C. $Y = 50X + 15$
    D. $X = 50Y + 15$
    E. $Y = 15X - 50$

14. If Lisa (#13) has been saving for 10 weeks, how much money does she have?
    A. $200
    B. $600
    C. $515
    D. $10
    E. $100

15. Katie started with the same amount as Lisa (#13) and saved $20 a week. How much will Katie have after 10 weeks?
    A. $520
    B. $70
    C. $300
    D. $200
    E. $250

# LESSON TEST 17

Circle your answer.

1. Which of the following represents three consecutive integers?
   A. N, N + 3, N + 5
   B. N, N + 1, N + 2
   C. N, N + 2, N + 4
   D. N, N + 2, N + 3
   E. N, 2N, 3N

2. Which of the following represents three consecutive odd integers?
   A. N, N + 3, N + 5
   B. N, N + 1, N + 2
   C. N, N + 2, N + 4
   D. N, N + 2, N + 3
   E. N, 2N, 3N

3. Which of the following represents three consecutive even integers?
   A. N, N + 3, N + 5
   B. N, N + 1, N + 2
   C. N, N + 2, N + 4
   D. N, N + 2, N + 3
   E. N, 2N, 3N

4. There are three consecutive integers, such that six times the first, minus two times the second, equals the third integer plus five. Which is the correct equation?
   A. 6(N + 1) − 2(N + 2) = 5(N + 3)
   B. 6N − 2(N + 1) = 5(N + 3)
   C. 6N − 2(N + 2) = 5(N + 4)
   D. 6N − 2(N + 1) = (N + 2) + 5
   E. 6N − 2(N + 2) = (N + 4) + 5

5. What is the first integer described in #4?
   A. 1
   B. 2
   C. 3
   D. 4
   E. 5

6. There are three consecutive integers, such that six times the first, plus five times the third, plus three times the third, equals 10 times the second, plus 10. What is the first integer?
   A. 1
   B. 2
   C. 3
   D. 4
   E. −1

7. What is the third integer described in #6?
   A. 2
   B. 3
   C. 4
   D. 5
   E. 6

8. There are three consecutive even integers, such that three times the first, plus the second, plus two, equals three times the third. What is the first integer?
   A. 2
   B. 4
   C. 8
   D. 10
   E. 12

ALGEBRA 1 LESSON TEST 17

LESSON TEST 17

9. What is the second integer described in #8?
    A. 3
    B. 4
    C. 5
    D. 6
    E. 10

10. There are three consecutive odd integers, such that 10 times the first, plus 10 times the second, equals 10 more than 10 times the third. What is the second integer?
    A. 7
    B. 4
    C. 3
    D. 5
    E. 9

11. What is the first integer described in #10?
    A. 3
    B. 4
    C. 5
    D. 6
    E. 7

12. There are three consecutive odd integers, such that three times the first, minus two times the second, plus 13, equals negative three times the sum of the first and the third. What is the first integer?
    A. −3
    B. −2
    C. 1
    D. 2
    E. 3

13. What is the third integer described in #12?
    A. −3
    B. −2
    C. 1
    D. 2
    E. 3

14. There are three consecutive integers, such that four times the first, plus two times the second, is equal to four times the third. What is the first integer?
    A. 1
    B. 2
    C. 3
    D. 4
    E. 5

15. What is the third integer described in #14?
    A. 3
    B. 4
    C. 5
    D. 2
    E. 6

# LESSON TEST 18

Circle your answer.

1. $(-6)^2$
   - A. 3
   - B. -12
   - C. 12
   - D. -36
   - E. 36

2. $-6^2$
   - A. 3
   - B. -12
   - C. 12
   - D. -36
   - E. 36

3. $R^2 \times R^4 =$
   - A. $R^6$
   - B. $R^8$
   - C. $R^{12}$
   - D. $1/R^4$
   - E. $R^1$

4. $R^4 \div R^2 =$
   - A. $1/R^4$
   - B. $R^2$
   - C. $R^8$
   - D. $R^{12}$
   - E. $R^4$

5. $R^8 \div R^2 =$
   - A. $R^8$
   - B. $R^{10}$
   - C. $R^4$
   - D. $R^{16}$
   - E. $R^6$

6. $A^{5X} \cdot A^{3X} =$
   - A. $A^{2X}$
   - B. $A^{8X}$
   - C. $A^2$
   - D. $AX^8$
   - E. $A^{615X}$

7. $C^4 C^3 D^2 D^1 =$
   - A. $C^{12}D^2$
   - B. $(CD)^{10}$
   - C. $C^7D^3$
   - D. $C^1D^1$
   - E. $C^6D^4$

8. Simplify $\sqrt{144}$
   - A. 4
   - B. 144
   - C. 12
   - D. 14
   - E. $12^2$

9. $4^8 \div 4^2 =$
   - A. $4^6$
   - B. $4^{-6}$
   - C. $4^{16}$
   - D. $4^{10}$
   - E. $4^{-10}$

10. $X^2 Y^3 X^4 Y =$
    - A. $X^6Y^3$
    - B. $X^8Y^3$
    - C. $X^6Y^4$
    - D. $XY^{10}$
    - E. $(XY)^{10}$

ALGEBRA 1 LESSON TEST 18

LESSON TEST 18

11. Simplify $-\sqrt{A^2}$
    A. $A^2$
    B. $-A$
    C. $A$
    D. $1$
    E. $-A^2$

12. Simplify $\sqrt{81B^2}$
    A. $-9B$
    B. $81B$
    C. $9B$
    D. $9B^2$
    E. $9\sqrt{B^2}$

13. Simplify $\sqrt{2^2 2^2}$
    A. 16
    B. 8
    C. 4
    D. 2
    E. $-4$

14. When dividing two numbers with the same base, the exponents are:
    A. added
    B. subtracted
    C. multiplied
    D. divided
    E. squared

15. When multiplying two numbers with the same base, the exponents are:
    A. added
    B. subtracted
    C. multiplied
    D. divided
    E. squared

# LESSON TEST 19

Circle your answer.

1. Rewrite on one line: $\dfrac{1}{X^{-3}}$
   - A. $X^{-3}$
   - B. $X^1$
   - C. $X^3$
   - D. $X^4$
   - E. $X^{-4}$

2. Rewrite on one line: $\dfrac{1}{X^3 X^4}$
   - A. $X^{-1}$
   - B. $X^1$
   - C. $X^7$
   - D. $X^{-7}$
   - E. $X^{-12}$

3. Write $X^{-4}$ without using negative exponents.
   - A. $X^8$
   - B. $X^{1/8}$
   - C. $X^1$
   - D. $X^0$
   - E. $1/X^4$

4. Write $5^{-5}$ without using negative exponents.
   - A. $5^5$
   - B. $5^{1/5}$
   - C. 25
   - D. $5^1$
   - E. $1/5^5$

5. $8^{-2} \cdot 8^{-2} =$
   - A. $8^{-4}$
   - B. $8^4$
   - C. $8^2$
   - D. 32
   - E. $16^{-2}$

6. $7^{-5} \div 7^3 =$
   - A. $7^{-2}$
   - B. $7^{-5/3}$
   - C. $7^{-8}$
   - D. $1^{-2}$
   - E. $7^{-15}$

7. $X^8 \div X^2 =$
   - A. $X^4$
   - B. $X^{16}$
   - C. $X^{-6}$
   - D. $X^{-4}$
   - E. $X^6$

8. Write $X^{-2} X^{-3}$ with positive exponents.
   - A. $1/X^5$
   - B. $X^5$
   - C. $X^6$
   - D. $1/X^6$
   - E. $X^1$

9. $X^0 =$
   - A. $X$
   - B. 1
   - C. 10
   - D. $0^X$
   - E. $1/X$

10. $X^{-2} Y^6 X^{-3} Y =$
    - A. $X^6 Y^6$
    - B. $X^5 Y^7$
    - C. $X^{-6} Y^5$
    - D. $(XY)^2$
    - E. $X^{-5} Y^7$

LESSON TEST 19

11. $A^{-1} A^{-8} B^7 B^2 =$
    A. $A^9 B^{-9}$
    B. $A^{-9} B^9$
    C. $A^8 B^{14}$
    D. $(AB)^{-81}$
    E. $A^{-7} B^5$

12. Simplify $\dfrac{B^4 B^2}{B^{-3}}$
    A. $B^3$
    B. $B^5$
    C. $B^{24}$
    D. $B^{-3}$
    E. $B^9$

13. Simplify $\dfrac{P^3 N^{-2}}{B^2 P^4}$
    A. $P^{-1} N^{-4}$
    B. $P^7 N^4$
    C. $P^1 N^4$
    D. $(PN)^{-5}$
    E. $P^{-5} N^7$

14. $\left(9^2\right)^5$
    A. $9^3$
    B. $9^7$
    C. $9^{-3}$
    D. $9^{10}$
    E. $9^{32}$

15. $\left(X^A\right)^B$
    A. $X^{A-B}$
    B. $X^{A+B}$
    C. $X^{AB}$
    D. $X^{B-A}$
    E. $X^{-AB}$

# LESSON TEST 20

Circle your answer.

1. $X^2 + 2X + 2$ is a:

   I. polynomial   II. trinomial
   III. binomial   IV. monomial

   A. I and II
   B. I and IV
   C. I only
   D. II only
   E. III only

2. $\quad X^2 + 3X + 2$
   $+ X^2 + 4X + 5$

   A. $X^2 + 7X + 7$
   B. $2X^2 + 7X + 3$
   C. $9X + 7$
   D. $2X^2 + 7X + 7$
   E. $2X^2 - X + 7$

3. $\quad X^2 + X + 10$
   $+ X^2 - 2X + 4$

   A. $2X^2 - X + 14$
   B. $X^2 - X + 14$
   C. $-X + 6$
   D. $2X^2 - 3X - 6$
   E. $2X^2 + X + 14$

4. $\quad X^2 + 8X + 6$
   $+ X^2 - 3X - 1$

   A. $X^2 + 5X + 5$
   B. $2X^2 - 5X - 5$
   C. $-11X + 7$
   D. $2X^2 + 11X + 7$
   E. $2X^2 + 5X + 5$

5. $\quad X^2 - 5X - 2$
   $+ X^2 - 4X - 3$

   A. $X^2 + 9X + 5$
   B. $9X + 5$
   C. $2X^2 - X - 1$
   D. $X^2 - 9X - 5$
   E. $2X^2 - 9X - 5$

6. What is the sum of $2X + 3$ and $4X - 5$?

   A. $6X^2 - 2$
   B. $6X + 2$
   C. $6X - 2$
   D. $6X + 8$
   E. $2X + 2$

7. What is the sum of $2X^2 - 9X + 5$ and $X^2 + 4X - 1$?

   A. $3X^2 + 5X + 4$
   B. $3X^2 - 5X + 4$
   C. $X^2 - 5X + 4$
   D. $3X^2 + 13X + 4$
   E. $3X^2 - 5X + 6$

8. $\quad 4X + 3$
   $\times\ X + 1$

   A. $5X^2 + 5X + 4$
   B. $11X + 3$
   C. $4X^2 + 7X + 3$
   D. $4X^2 + 7X + 4$
   E. $4X^2 + X + 3$

ALGEBRA 1 LESSON TEST 20

LESSON TEST 20

9.  $X + 3$
    $\times X + 2$

    A. $X^2 + 6X + 5$
    B. $X^2 + 5X + 6$
    C. $2X^2 + 5X + 6$
    D. $X^2 + X + 5$
    E. $X^2 + X + 6$

10. The product of $X + 4$ and $X - 2$ is:
    A. $X^2 + 2X - 8$
    B. $X^2 - 2X - 8$
    C. $2X^2 + 6X - 8$
    D. $X^2 - 6X - 8$
    E. $X^2 - 2X + 8$

11. Multiply $X + 1$ and $X + 5$.
    A. $X^2 + 5X + 6$
    B. $X^2 + 6X - 5$
    C. $X^2 + 6X + 5$
    D. $X^2 + 5X + 4$
    E. $2X^2 + 6X + 5$

12. Multiply $X - 3$ and $X - 6$.
    A. $X^2 + 9X - 18$
    B. $X^2 + 9X + 18$
    C. $2X^2 - 9X + 18$
    D. $X^2 - 9X + 18$
    E. $X^2 - 18X - 9$

13. If $7X + 1$ and $X + 2$ are multiplied, the first term of the answer would be:
    A. $X^2$
    B. $7X^2$
    C. $14X^2$
    D. $2X^2$
    E. $7X$

14. If $2X + 4$ and $X + 5$ are multiplied, the first term of the answer would be:
    A. $3X^2$
    B. $2X^2$
    C. $10X^2$
    D. $8X^2$
    E. $20X^2$

15. When we multiply 2 binomials, the result is a(n):
    A. binomial
    B. trinomial
    C. monomial
    D. integer
    E. inequality

# LESSON TEST 21

Circle your answer.

1. If (X + A) is multiplied times (X + B), the final term of the resulting trinomial will be:
   A. $X^2$
   B. (A + B)X
   C. BX
   D. AX
   E. AB

2. If (X + A) is multiplied times (X + B), the middle term of the resulting trinomial will be:
   A. $X^2$
   B. (A + B)X
   C. BX
   D. AX
   E. AB

3. The factors of $X^2 + 3X + 2$ are:
   A. (X + 3)(X + 2)
   B. (X + 1)(X + 2)
   C. X(X + 2)
   D. (X + 5)(X + 2)
   E. (X − 1)(X + 2)

4. The factors of $X^2 + 8X + 15$ are:
   A. (X + 2)(X + 4)
   B. (X + 1)(X + 8)
   C. (X + 10)(X + 5)
   D. (X + 7)(X + 8)
   E. (X + 3)(X + 5)

5. The factors of $X^2 + 12X + 36$ are:
   A. (X + 3)(X + 4)
   B. (X + 6)(X + 6)
   C. (X + 6)(X + 2)
   D. (X + 18)(X + 18)
   E. (X − 6)(X + 6)

6. The factors of $X^2 + 12X + 20$ are:
   A. (X + 12)(X − 20)
   B. (X + 2)(X + 10)
   C. X(X + 20)
   D. (X + 5)(X + 4)
   E. (X + 12)(X + 20)

7. The factors of $X^2 + 11X + 24$ are:
   A. (X + 4)(X + 6)
   B. (X + 2)(X + 12)
   C. (X + 3)(X + 8)
   D. (X + 1)(X + 24)
   E. (X + 5)(X + 6)

8. The factors of $X^2 + 6X + 5$ are:
   A. (X + 2)(X + 3)
   B. (X + 1)(X + 6)
   C. X(X + 6)
   D. (X + 1)(X + 5)
   E. (X + 5)(X + 6)

9. The factors of $X^2 + 14X + 49$ are:
   A. (X + 7)(X + 7)
   B. (X + 1)(X + 49)
   C. X(X + 7)
   D. (X + 2)(X + 7)
   E. (X + 1)(X + 14)

10. The factors of $X^2 + 11X + 10$ are:
    A. (X + 2)(X + 5)
    B. (X + 1)(X + 10)
    C. X(X + 10)
    D. (X + 1)(X + 11)
    E. (X + 5)(X + 5)

LESSON TEST 21

11. $(A + B)(A + B)$ is equal to:
    A. $A^2 + BA + B^2$
    B. $A^2 + 2BA + AB^2$
    C. $A^2 + 2BA + (AB)^2$
    D. $A^2 + 2BA + B^2$
    E. $A^2 + A + B + B^2$

12. $(X + BY)(X + BY)$ is equal to:
    A. $X^2 + 2BYX + BY^2$
    B. $X^2 + BYX + (BY)^2$
    C. $X^2 + 2BY + (BY)^2$
    D. $X^2 + 2BY + BY^2$
    E. $X^2 + 2BYX + (BY)^2$

13. What are the factors of $X^2 + (R + T)X + RT$?
    A. $(X + X)(X + T)$
    B. $(RX)(TX)$
    C. $(X + R)(X + T)$
    D. $X(R + T)$
    E. $(R + T)(R + T)$

14. What are the factors of $X^2 + 2RX + R^2$?
    A. $(X + 2)(X + 2R)$
    B. $(X + R)(X + R)$
    C. $(X + 2R)(X + 2R)$
    D. $X(RX + R)$
    E. $(R + S)(X + R)$

15. Fill in the blank.

    The numbers that are added to get the coefficient of the middle term are the _____ of the last term.

    A. exponents
    B. factors
    C. inverse
    D. addends
    E. products

# LESSON TEST 22

Circle your answer.

1. If (2X + A) is multiplied times (X + A), the final term of the resulting trinomial will be:
    A. $2X^2$
    B. $A^2$
    C. $2A^2$
    D. 2XA
    E. 3XA

2. If (2X + A) is multiplied times (X + A), the second term of the resulting trinomial will be:
    A. $2X^2$
    B. $A^2$
    C. $2A^2$
    D. 2XA
    E. 3AX

3. The factors of $2X^2 + 5X + 2$ are:
    A. (X + 1)(X + 2)
    B. (2X + 1)(X + 2)
    C. 2(X + 2)
    D. (2X + 5)(X + 1)
    E. (2X + 1)(2X + 2)

4. The factors of $3X^2 + 14X + 8$ are:
    A. (3X + 2)(3X + 4)
    B. (3X + 4)(X + 4)
    C. (X + 2)(X + 4)
    D. (3X + 2)(X + 7)
    E. (3X + 2)(X + 4)

5. The factors of $2X^2 + 9X + 9$ are:
    A. (X + 3)(X + 3)
    B. (2X + 3)(2X + 3)
    C. (2X + 6)(X + 3)
    D. (2X + 3)(X + 3)
    E. (2X + 9)(X + 1)

6. One way $4X^2 + 10X + 6$ may be factored is:
    A. (4X + 6)(X + 1)
    B. (2X + 6)(2X + 1)
    C. (X + 6)(X + 1)
    D. (4X + 3)(X + 2)
    E. (4X + 6)(4X + 1)

7. The factors of $2X^2 + 3X + 1$ are:
    A. (X + 2)(X + 1)
    B. (2X + 1)(X + 1)
    C. (X + 1)(X + 3)
    D. (2X + 1)(2X + 1)
    E. (2X + 3)(X + 1)

8. The factors of $3A^2 + 10A + 8$ are:
    A. (3A + 4)(A + 2)
    B. (3A + 4)(A + 4)
    C. (3A + 4)(A + 1)
    D. (3A + 4)(3A + 2)
    E. 3(A + 2)(A + 5)

9. $2Y^2 + 12Y + 18$ may be factored as:
    A. (Y + 6)(Y + 3)
    B. (2Y + 9)(Y + 2)
    C. (Y + 3)(Y + 4)
    D. (2Y + 6)(Y + 2)
    E. (2Y + 6)(Y + 3)

10. The factors of $5B^2 + 12B + 4$ are:
    A. (5B + 1)(B + 4)
    B. (B + 2)(B + 2)
    C. (5B + 6)(B + 2)
    D. 5(B + 2)(B + 2)
    E. (5B + 2)(B + 2)

LESSON TEST 22

11. $(2A + B)(A + B)$ is equal to:
    A. $A^2 + 2AB + B^2$
    B. $A^2 + 2AB + AB^2$
    C. $2A^2 + (A + B) + B^2$
    D. $2A^2 + 3AB + B^2$
    E. $A^2 + 2A + 2B + B^2$

12. $(2A + B)(A + C)$ is equal to:
    A. $A^2 + ABC + BC$
    B. $2A^2 + (2C + B)A + BC$
    C. $2A^2 + (C + B)A + BC$
    D. $2A^2 + BC + C^2$
    E. $2A^2 + 2(C + B)A + BC$

13. What are the factors of $3X^2 + 4RX + R^2$?
    A. $(3X + R)(X + R)$
    B. $(3X + 2)(X + 2)$
    C. $(3X + 4)(X + R)$
    D. $(X + R)(X + 4R)$
    E. $(3X + R)(4X + R)$

14. What are the factors of $2A^2 + 7AB + 3B^2$?
    A. $(2A + 3B)(A + B)$
    B. $(A + 3)(A + 4B)$
    C. $(2A + 3B)(A + 4B)$
    D. $(2A + B)(A + 6B)$
    E. $(2A + B)(A + 3B)$

15. Which terms of the polynomial are affected by the coefficient "5" in the product of $(5X + Y)(X + Y)$?
    A. first only
    B. second only
    C. third only
    D. first and second
    E. first and third

# LESSON TEST 23

Circle your answer.

1. One key to determining the signs of the factors of a polynomial is:
   A. the first term
   B. GCF
   C. the last term
   D. the exponent
   E. the value of X

2. If (X − A) is multiplied times (X − B), and A and B are positive numbers, what will be the signs of the last 2 terms of the resulting trinomial?
   A. both positive
   B. both negative
   C. first positive, second negative
   D. first negative, second positive
   E. cannot be determined

3. (X − 2)(X − 3) is equal to:
   A. $X^2 - 5X + 6$
   B. $X^2 - 5X - 6$
   C. $X^2 + 5X + 6$
   D. $X^2 + 5X - 6$
   E. $X^2 - X + 6$

4. (X − 2)(X + 3) is equal to:
   A. $X^2 - 5X + 6$
   B. $X^2 - 5X - 6$
   C. $X^2 + X + 6$
   D. $X^2 + X - 6$
   E. $X^2 - X + 6$

5. (X + 2)(X − 3) is equal to:
   A. $X^2 - 6X + 5$
   B. $X^2 - 5X - 6$
   C. $X^2 - X - 6$
   D. $X^2 + X - 6$
   E. $X^2 - 5X + 6$

6. The factors of $X^2 + X - 2$ are:
   A. (X + 1)(X − 2)
   B. (X − 1)(X + 2)
   C. (X + 1)(X + 2)
   D. (X + 3)(X − 1)
   E. (X − 1)(X + 3)

7. The factors of $X^2 - 3X - 4$ are:
   A. X(3X − 4)
   B. (X + 2)(X − 2)
   C. (X − 3)(X − 4)
   D. (X + 4)(X − 1)
   E. (X − 4)(X + 1)

8. The factors of $X^2 - 5X + 6$ are:
   A. (X + 3)(X − 2)
   B. (X − 3)(X − 2)
   C. (X + 2)(X − 3)
   D. X(5X + 6)
   E. (X − 1)(X + 6)

LESSON TEST 23

9. The factors of $A^2 - A - 12$ are:
    A. $(A + 3)(A - 4)$
    B. $(A - 3)(A + 4)$
    C. $A(A - 12)$
    D. $(A - 3)(A - 4)$
    E. $(A - 1)(A + 12)$

10. The factors of $A^2 + A - 12$ are:
    A. $(A + 3)(A - 4)$
    B. $(A - 3)(A + 4)$
    C. $A(A - 12)$
    D. $(A - 3)(A - 4)$
    E. $(A - 1)(A + 12)$

11. $(X - Y)(X - Y)$ is equal to:
    A. $X^2 - XY + Y^2$
    B. $X^2 + XY - Y^2$
    C. $X^2 + Y^2$
    D. $X^2 - Y^2$
    E. $X^2 - 2XY + Y^2$

12. $(X + Y)(X - Y)$ is equal to:
    A. $X^2 - XY + Y^2$
    B. $X^2 + XY - Y^2$
    C. $X^2 + Y^2$
    D. $X^2 - Y^2$
    E. $X^2 - 2XY + Y^2$

13. What are the factors of $X^2 - 2RX + R^2$?
    A. $(X + 2)(X + 2R)$
    B. $(X + R)(X - R)$
    C. $(X - R)(X + R)$
    D. $(X - R)(X - R)$
    E. $(X + 2)(X - 2R)$

14. If the final term of a trinomial is positive, describe the second terms of each factor.
    A. Both must be positive.
    B. Both must be negative.
    C. The first must be negative.
    D. The second must be negative.
    E. They are either both negative or both positive.

15. If the first term of a trinomial has no coefficients, the middle term takes the sign of:
    A. the second term of the first factor
    B. the second term of the last factor
    C. the second term of either factor with the smaller value
    D. the second term of either factor with the larger value
    E. The sign of the middle term cannot be determined.

UNIT TEST  **Lessons 12-23** (100 points possible)

I. Evaluate. (5 points each)

1. $5^2 \cdot 5^3$

2. $(-3)^3$

3. $\left(2^{-2}\right)^3$

4. $3^{10} \div 3^2$

5. $A^2B^3AB^4$

6. $(3X + 2)(X - 1)$

II. Sally has 16 coins. All are either nickels or dimes. The value of her coins is $1.10. How many dimes does she have? (10 points)

UNIT TEST II

III. Solve the following problems using graphing, substitution, or elimination. Use a different method for each problem. Show your work and label the method used. Use graph if desired. (10 points each)

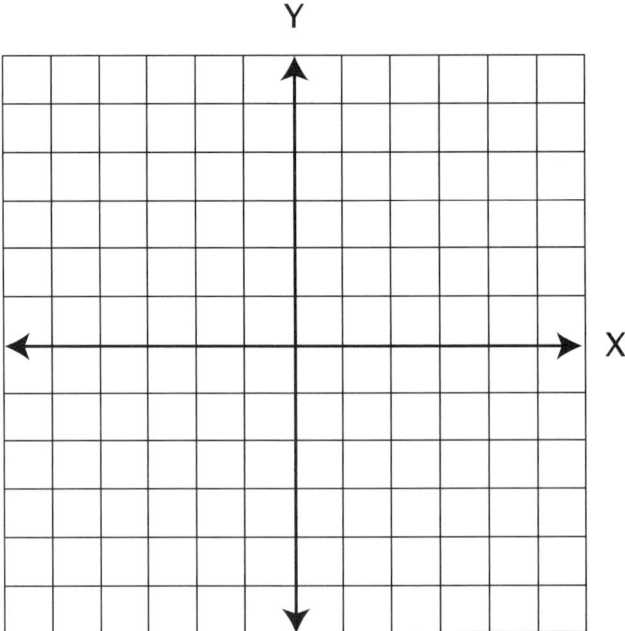

1.  2X + 1 = Y
    2Y = 6

2.  Y − 3 = X + 2
    2X − 1 = Y

IV. Find three consecutive odd integers where two times the first integer, plus one, equals the third integer. (8 points)

V. Factor. (8 points each)

1. $2X^2 + 28$

2. $2X^2 + 8X + 6$

3. $3X^2 + 19X + 20$

VI. Graph the inequality: $2Y \leq 4X - 8$. (8 points)

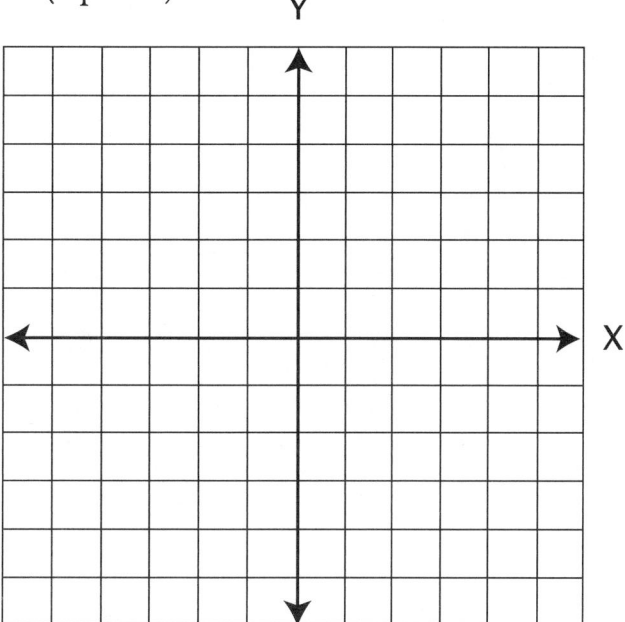

# LESSON TEST 24

Circle your answer.

1. Which of these is a square root of $X^2 + 6X + 9$?
   A. $X + 3$
   B. $(X + 3)^2$
   C. $X + 2$
   D. $X + 9$
   E. $(X + 6)^2$

2. Which of these is a square root of $X^2 + 8X + 16$?
   A. $(X + 4)^2$
   B. $X + 2$
   C. $X + 8$
   D. $X - 4$
   E. $X + 4$

3. Which of these is a square root of $X^2 + 2X + 1$?
   A. $X - 1$
   B. $X + 1$
   C. $(X + 1)^2$
   D. $X + 2$
   E. $(X + 2)^2$

4. $\sqrt{X^2 + 12X + 36} =$
   A. $(X + 3)^2$
   B. $X + 3$
   C. $X + 6$
   D. $X - 6$
   E. $(X + 6)^2$

5. $\sqrt{X^2 + 4X + 4} =$
   A. $(X + 2)^2$
   B. $X + 2$
   C. $X + 4$
   D. $X - 2$
   E. $(X + 4)^2$

6. $\sqrt{X^2 + 10X + 25} =$
   A. $(X + 2)^2$
   B. $X - 5$
   C. $X + 10$
   D. $(X + 5)^2$
   E. $X + 5$

7. Divide $X^2 + 3X + 2$ by $X + 2$.
   A. $X + 1$
   B. $X + 2$
   C. $X - 1$
   D. $X^2 + 1$
   E. $X^2 - 1$

8. Divide $X^2 + 9X + 20$ by $X + 3$.
   A. $X + 2 \, r \, 6$
   B. $X + 8$
   C. $X + 6$
   D. $X + 6 \, r \, 2$
   E. $X + 12$

LESSON TEST 24

9. Divide $X^2 + 4X - 5$ by $X + 4$.
   A. $X - 5$
   B. $X + 5$
   C. $X - 5$ r 1
   D. $X$ r 5
   E. $X$ r -5

10. $X + 3 \overline{) X^2 - 4X^1 - 21} =$
    A. $X + 42$
    B. $X - 7$
    C. $X + 7$
    D. $X - 42$
    E. $X - 3$

11. $X + 5 \overline{) X^2 + 8X^1 + 15} =$
    A. $X + 5$
    B. $X - 5$
    C. $X + 3$
    D. $X - 3$
    E. $X + 4$

12. $X + 2 \overline{) X^2 + 6X + 10} =$
    A. $X + 5$
    B. $X + 4$ r 2
    C. $X + 2$ r 4
    D. $X - 5$
    E. $X + 3$

13. What is $(X + 5)^2$?
    A. $X^2 + 10X + 25$
    B. $X^2 + 5X + 25$
    C. $2X + 10$
    D. $X^2 - 10X + 25$
    E. $X^2 + 25$

14. What is the square of $X + 7$?
    A. $X^2 + 40X + 7$
    B. $X^2 + 14X - 49$
    C. $X^2 + 49$
    D. $X^2 + 14X + 49$
    E. $X^2 + 7$

15. What is the product of $2X + 7$ and $X + 5$?
    A. $2X + 17X + 35$
    B. $2X^2 + 17X + 35$
    C. $2X^2 + 12X + 35$
    D. $X^2 + 12X + 35$
    E. $2X^2 + 35$

# LESSON TEST

## 25

★ Circle your answer.

1. $(X + B)(X - B)$ is equal to:
    A. $X^2 - 2B$
    B. $X^2 + B^2$
    C. $X^2 + BX + B^2$
    D. $X^2 + 2X + B^2$
    E. $X^2 - B^2$

2. What are the factors of $R^2 - 16$?
    A. $(R - 2)(R + 8)$
    B. $R(R - 16)$
    C. $(R + 4)(R - 4)$
    D. $(R - 4)(R - 4)$
    E. $(R - 2)(R - 8)$

3. What are the factors of $S^2 - 25$?
    A. $(S - 5)(S - 5)$
    B. $(S - 30)(S - 5)$
    C. $(S - 20)(S - 5)$
    D. $(S + 5)(S - 5)$
    E. $S(S - 25)$

4. $(R + T)(R - T)$ is equal to:
    A. $R^2 - T^2$
    B. $R^2 + T^2$
    C. $R^2 + 2RT$
    D. $R^2 + TR + T^2$
    E. $R^2 - TR + T^2$

5. What are the factors of $X^2 - 36$?
    A. $(X - 6)(X + 6)$
    B. $(X - 6)(X - 6)$
    C. $(X + 6)(X + 6)$
    D. $(X - 4)(X + 9)$
    E. none of the above

6. What are the factors of $X^2 + 36$?
    A. $(X - 6)(X + 6)$
    B. $(X - 6)(X - 6)$
    C. $(X + 6)(X + 6)$
    D. $(X - 4)(X + 9)$
    E. none of the above

7. What are the factors of $X^2 + 100$?
    A. $(X + 10)(X + 10)$
    B. $(X - 10)(X - 10)$
    C. $(X + 10)(X - 10)$
    D. $(X + 50)(X + 2)$
    E. none of the above

8. What are the factors of $X^2 - 100$?
    A. $(X + 10)(X + 10)$
    B. $(X - 10)(X - 10)$
    C. $(X + 10)(X - 10)$
    D. $(X + 50)(X + 2)$
    E. none of the above

LESSON TEST 25

9. (X + 8)(X − 8) is equal to :
   A. X² − 16
   B. X² + 64
   C. X² − 64
   D. X² − 16X + 64
   E. X² + 16X − 64

10. (X + 3)(X − 3) is equal to:
    A. X² − 9
    B. X² + 9
    C. X² − 6
    D. X² − 6X + 9
    E. X² + 6X − 9

11. (X + 7)(X − 7) is equal to:
    A. X² − 14
    B. X² + 14
    C. X² + 49
    D. X² − 49
    E. X² + 14X − 49

12. In order to solve a problem with oriental squares, the numbers in the units places must:
    A. be the same.
    B. be different.
    C. add to 10.
    D. be fives.
    E. be zeros.

13. In order to solve a problem with oriental squares, the numbers in the tens places must:
    A. be the same.
    B. be different.
    C. add to 10.
    D. add to five.
    E. be ones.

14. Solve 45 x 45 using oriental squares.
    A. 1,625
    B. 2,025
    C. 225
    D. 900
    E. 1,325

15. Solve 63 x 67 using oriental squares.
    A. 3,621
    B. 3,610
    C. 4,510
    D. 4,221
    E. 130

# LESSON TEST 26

Circle your answer.

1. $X^4 - 81$ is the same as:

   A. $(X^2+9)(X^2-3)$

   B. $(X^2+9)(X-3)(X-3)$

   C. $X^2(X^2-81)$

   D. $(X^2+9)(X-3)(X+3)$

   E. $(X+9)(X-9)$

2. $X^4 - 9$ is the same as:

   A. $(X^2+9)(X^2-9)$
   B. $(X+3)(X-3)$
   C. $X^2(X^2-3)$
   D. $(X^2+3)(X^2-3)$
   E. $(X^2+3)(X-1)(X+3)$

3. $A^4 - 16$ is the same as:

   A. $(A^2+8)(A^2-8)$

   B. $(A^2+8)(A+4)(A-4)$

   C. $(A+2)(A-2)$

   D. $(A^2-4)(A+2)(A-2)$

   E. $(A^2+4)(A+2)(A-2)$

4. To factor $4X^4 - 4X^2$, you could first divide each term by any of the following except:

   A. 4
   B. X
   C. $X^2$
   D. $4X^4$
   E. $4X^2$

5. To factor $5X^3 - 5X^2 - 30X$, you could first divide each term by which of the following:

   A. $X^3$
   B. $5X$
   C. $X^2$
   D. $5X^3$
   E. $5X^2$

6. Write $B^4 - 1,000$ as the difference of two squares.

   A. $(B+100)(B-100)$
   B. $(B^2+10)(B^2-10)$
   C. $(B+10)(B-10)$
   D. $(B^2+100)(B^2-100)$
   E. none of the above

7. Factored completely, $B^4 - 10,000$ is:

   A. $(B^2+100)(B^2-100)$

   B. $(B^2+10)(B+10)(B-10)$

   C. $(B+10)(B-10)$

   D. $(B^2+100)(B+10)(B-10)$

   E. none of the above

LESSON TEST 26

8. $X^4 - Y^4$ can be partially factored as:
   A. $(X + Y)(X - Y)$
   B. $(X^2 + Y^2)(X + Y)(X - Y)$
   C. $(X^2 + Y^2)(X^2 - Y^2)$
   D. $(X^2 - Y^2)(X^2 - Y^2)$
   E. none of the above

9. Factored completely, $X^4 - Y^4$ is:
   A. $(X + Y)(X - Y)$
   B. $(X^2 + Y^2)(X + Y)(X - Y)$
   C. $(X^2 + Y^2)(X^2 - Y^2)$
   D. $(X^2 - Y^2)(X^2 - Y^2)$
   E. none of the above

10. Factor completely:
    $2X^3 + 16X^2 + 24X$.
    A. $(X + 2)(X + 6)$
    B. $(2X^2 + 4)(X + 6)$
    C. $2X(X + 2)(X - 6)$
    D. $2X(X + 2)(X + 6)$
    E. $2X^2(X + 2)(X - 6)$

11. Factor completely: $4X^3 - 64X$.
    A. $(2X + 8)(2X - 8)$
    B. $(2X^2 + 8)(X - 8)$
    C. $4X(X + 2)(X - 2)$
    D. $4X^2(X + 4)(X - 4)$
    E. $4X(X + 4)(X - 4)$

12. Factor completely:
    $3X^3 - 12X^2 - 15X$.
    A. $3X(X - 5)(X + 1)$
    B. $(3X^2 - 5)(X + 1)$
    C. $3X(X + 5)(X - 1)$
    D. $3X(X + 5)(X + 1)$
    E. $3X(X - 5)(X + 3)$

13. Factor completely: $8X^3 - 72X$.
    A. $8X(X - 8)(X + 9)$
    B. $(4X^2 - 4)(2X + 8)$
    C. $8X(X + 3)(X - 3)$
    D. $4X^2(X - 9)$
    E. $8X(X - 9)(X + 9)$

14. How many hours will it take to travel 480 miles at 60 mph?
    A. 6
    B. 8
    C. 48
    D. 480
    E. 28,800

15. If you travel at 65 mph for five hours, how many miles will you travel?
    A. 13
    B. 11
    C. 305
    D. 325
    E. 1.5

LESSON TEST

# 27

Circle your answer.

1. In order to solve
   $X^2 + 2X + 4 = 2$, first:
   A. divide each term by 2
   B. factor $X^2 + 6X + 4$
   C. subtract 2 from each side
   D. try a possible value of X
   E. write $X^2 + 6X + 4 = 0$

2. In the expression $(X + 7)(X - 2) = 0$, which cannot be true?
   A. $X + 7 = 0$
   B. $X - 2 = 0$
   C. $X = 2$
   D. $X = -7$
   E. $X = 7$

3. In order to solve
   $2X^2 + 6X + 4 = 0$, first:
   A. divide each term by 2
   B. factor $2X^2 + 6X + 4$
   C. subtract 4 from each side
   D. try a possible value of X
   E. write $2X^2 + 6X = 4$

4. In order to solve $X^3 - 16X = 0$, first:
   A. write $(X + 4)(X - 4) = 0$
   B. factor out X
   C. subtract X from each side
   D. make $X = 0$
   E. write $X - 16 = 0$

5. In the expression
   $2X(X + 1)(X - 3) = 0$,
   which cannot be true?
   A. $X = 1$
   B. $X = -1$
   C. $X = 3$
   D. $2X = 0$
   E. $X - 3 = 0$

6. Given $X^2 + 11X + 30 = 0$, one value of X is:
   A. 5
   B. 6
   C. -5
   D. 2
   E. 15

7. Given $2X^2 + 7X + 6 = 0$, one value of X is:
   A. -2
   B. 2
   C. 3
   D. 1
   E. -1

8. Given $2X^2 - 7X + 6 = 0$, one value of X is:
   A. 3
   B. -3
   C. -2
   D. 7
   E. 3/2

LESSON TEST 27

9. Solve for X: $X^2 + 9X + 40 = 20$.
   A. 4, 5
   B. −4, −5
   C. 3, −3
   D. −3, −5
   E. −4, 5

10. Solve for X: $3X^2 − 3X − 18 = 0$.
    A. 2, 9
    B. −3, 6
    C. 3, −6
    D. −2, 3
    E. −3, 2

11. Solve for X: $X^2 − 8X + 16 = 1$.
    A. −4, −4
    B. −5, −3
    C. 5, 3
    D. 4, 4
    E. −5, 3

12. The solution set for $2X^2 − 2X − 4 = 20$ is:
    A. 2, 1
    B. −2, −1
    C. −3, −4
    D. 3, 4
    E. −3, 4

13. The solution set for $3X^2 + 9X = 12$ is:
    A. −2, 1
    B. 2, 2
    C. 3, −4
    D. −4, 1
    E. 4, −1

14. The solution set for $X^2 − 10X + 25 = 0$ is:
    A. −5, −5
    B. 5, 5
    C. 2, 5
    D. −2, 5
    E. 2, −5

15. In the equation $X^2 + (R + S)X + RS = 0$, what is one value of X if R and S ≠ 0?
    A. −S
    B. 0
    C. 3RS
    D. R + S
    E. S

# LESSON TEST 28

Circle your answer.

1. Which unit multiplier should be used to change feet to inches?

    A. $\dfrac{1 \text{ ft}}{12 \text{ in}}$

    B. $\dfrac{12 \text{ in}}{1 \text{ ft}}$

    C. $\dfrac{12 \text{ ft}}{1 \text{ in}}$

    D. $\dfrac{1 \text{ in}}{12 \text{ ft}}$

    E. $\dfrac{12 \text{ in}}{12 \text{ in}}$

2. Which unit multiplier should be used to change quarts to gallons?

    A. $\dfrac{4 \text{ qt}}{1 \text{ gal}}$

    B. $\dfrac{8 \text{ qt}}{1 \text{ gal}}$

    C. $\dfrac{1 \text{ gal}}{8 \text{ qt}}$

    D. $\dfrac{1 \text{ gal}}{4 \text{ qt}}$

    E. $\dfrac{2 \text{ pt}}{4 \text{ qt}}$

3. Which unit multiplier should be used to change feet to yards?

    A. $\dfrac{1 \text{ yd}}{36 \text{ in}}$

    B. $\dfrac{3 \text{ ft}}{3 \text{ ft}}$

    C. $\dfrac{1 \text{ yd}}{3 \text{ ft}}$

    D. $\dfrac{3 \text{ ft}}{1 \text{ yd}}$

    E. $\dfrac{1 \text{ yd}}{2 \text{ ft}}$

4. Which unit multiplier should be used to change pounds to ounces?

    A. $\dfrac{16 \text{ oz}}{1 \text{ lb}}$

    B. $\dfrac{1 \text{ lb}}{16 \text{ oz}}$

    C. $\dfrac{1 \text{ lb}}{8 \text{ oz}}$

    D. $\dfrac{8 \text{ oz}}{1 \text{ lb}}$

    E. $\dfrac{8 \text{ oz}}{8 \text{ oz}}$

LESSON TEST 28

5. Which unit multiplier should be used to change pounds to tons?

   A. $\dfrac{1,000 \text{ lb}}{1 \text{ ton}}$

   B. $\dfrac{1 \text{ ton}}{1,000 \text{ lb}}$

   C. $\dfrac{1 \text{ ton}}{1 \text{ ton}}$

   D. $\dfrac{1 \text{ ton}}{2,000 \text{ lb}}$

   E. $\dfrac{2,000 \text{ lb}}{1 \text{ ton}}$

6. Which unit multiplier should be used to change pints to quarts?

   A. $\dfrac{2 \text{ pt}}{1 \text{ qt}}$

   B. $\dfrac{1 \text{ qt}}{2 \text{ pt}}$

   C. $\dfrac{2 \text{ qt}}{1 \text{ pt}}$

   D. $\dfrac{2 \text{ pt}}{2 \text{ pt}}$

   E. $\dfrac{2 \text{ qt}}{2 \text{ pt}}$

7. How many feet is 48 inches?
   A. 12
   B. 4
   C. 576
   D. 1 1/3
   E. 3

8. How many quarts is 16 gallons?
   A. 8
   B. 32
   C. 64
   D. 128
   E. 6

9. How many ounces are in 10 pounds?
   A. 160
   B. 1
   C. 16
   D. 106
   E. 140

10. How many pints are in six quarts?
    A. 1 1/2
    B. 24
    C. 3
    D. 12
    E. 2

11. Eight thousand pounds equals how many tons?
    A. 1,000
    B. 4
    C. 16,000
    D. 8
    E. 2

LESSON TEST 28

12. Eighty ounces equals how many pounds?
    A. 1,280
    B. 1,120
    C. 5
    D. 10
    E. 6

13. Six yards equals how many inches?
    A. 2
    B. 72
    C. 6
    D. 216
    E. 180

14. When using unit multipliers, the name of the measure in the numerator should be the same as:
    A. the name in the denominator
    B. the name of the given amount
    C. the name of the unit in the desired answer
    D. the larger of the units of measure given
    E. the smaller of the units of measure given

15. The term used to multiply by when converting measures should always be:
    A. less than the given measure
    B. greater than the given measure
    C. greater than one
    D. less than one
    E. equal to one

ALGEBRA 1

LESSON TEST 28

LESSON TEST

# 29

Circle your answer.

1. How many unit multipliers are required to change square feet to square inches?
   A. 1
   B. 2
   C. 3
   D. 12
   E. 144

2. How many unit multipliers are required to change cubic yards to cubic feet?
   A. 1
   B. 2
   C. 3
   D. 9
   E. 12

3. Which unit multipliers should be used to change two miles to inches?

   I. $\dfrac{1 \text{ ft}}{2 \text{ in}}$

   II. $\dfrac{12 \text{ in}}{1 \text{ ft}}$

   III. $\dfrac{1 \text{ mi}}{5{,}280 \text{ ft}}$

   IV. $\dfrac{5{,}280 \text{ ft}}{1 \text{ mi}}$

   V. $\dfrac{12 \text{ in}}{5{,}280 \text{ ft}}$

   A. I, II
   B. III, IV
   C. II, IV
   D. III, V
   E. V only

4. Which unit multipliers should be used to change 16 pints to gallons?

   I. $\dfrac{4 \text{ qt}}{1 \text{ gal}}$

   II. $\dfrac{2 \text{ pt}}{1 \text{ qt}}$

   III. $\dfrac{1 \text{ gal}}{4 \text{ pt}}$

   IV. $\dfrac{1 \text{ qt}}{2 \text{ pt}}$

   V. $\dfrac{8 \text{ qt}}{1 \text{ gal}}$

   A. I, II
   B. III, IV
   C. II, V
   D. III, V
   E. V only

5. How many square feet are there in two square yards of carpet?
   A. 6
   B. 10
   C. 24
   D. 36
   E. 18

6. How many cubic feet will you receive if you order six yards of concrete?
   A. 9
   B. 18
   C. 27
   D. 162
   E. 54

ALGEBRA 1 LESSON TEST 29

LESSON TEST 29

7. How many square inches are in eight square yards?
    A. 1,152
    B. 288
    C. 10,368
    D. 72
    E. 2,304

8. How many acres is 87,120 square feet?
    A. 1
    B. 2
    C. 3
    D. 4
    E. 5

9. Sam stacked a pile of wood that was four feet deep, four feet high, and 16 feet long. How many cords of wood does he have?
    A. 1
    B. 1 1/2
    C. 2
    D. 2 1/2
    E. 3

10. Six cubic meters equal how many cubic centimeters?
    A. 60
    B. 600
    C. 6,000
    D. 60,000
    E. 6,000,000

For #11–15, answer:

   A  if the quantity in column A is greater.
   B  if the quantity in column B is greater.
   C  if the two quantities are equal.
   D  if the relationship cannot be determined from the information given.

Write your answer in the blank.

| # | A | B | |
|---|---|---|---|
| 11. | $3 \text{ ft}^2$ | $1 \text{ yd}^2$ | _____ |
| 12. | $9 \text{ ft}^2$ | $1 \text{ yd}^2$ | _____ |
| 13. | $1 \text{ mi}^2$ | $1 \text{ acre}$ | _____ |
| 14. | $X \text{ mi}^2$ | $Y \text{ in}^2$ | _____ |
| 15. | $215 \text{ ft}^3$ | $8 \text{ yd}^3$ | _____ |

# LESSON TEST 30

Use this information as needed to answer the questions. The equivalents are approximate.

**U. S. Customary to Metric**
1 inch ≈ 2.5 centimeters
1 yard ≈ .9 meters
1 mile ≈ 1.6 kilometers
1 ounce ≈ 28 grams
1 pound ≈ .45 kilograms
1 quart ≈ .95 liters

**Metric to U. S. Customary**
1 centimeter ≈ .4 inches
1 meter ≈ 1.1 yards
1 kilometer ≈ .62 miles
1 gram ≈ .035 ounces
1 kilogram ≈ 2.2 pounds
1 liter ≈ 1.06 quarts

Circle your answer.

1. Five kilometers is how many miles?
   A. 8
   B. 1.61
   C. 8.06
   D. 3.1
   E. 500

2. Change six kilograms to pounds.
   A. .37
   B. 13.2
   C. 2.7
   D. .08
   E. 2.73

3. How many meters is three yards?
   A. 2.7
   B. 3.33
   C. 3.2
   D. .37
   E. 9

4. How many kilometers in six miles?
   A. 3.75
   B. 9.6
   C. 3.72
   D. 6,000
   E. 90

5. Eight pounds is how many kilograms?
   A. 12.8
   B. 17.78
   C. 4.96
   D. 3.6
   E. 80

6. Six meters equals:
   A. 5.45 yd
   B. 6.6 ft
   C. 5.4 yd
   D. 5.45 ft
   E. 6.6 yd

7. Nine centimeters is how many inches?
   A. 3.6
   B. .044
   C. 22.5
   D. .277
   E. 1.6

LESSON TEST 30

8. Nine inches is how many centimeters?
   A. 3.6
   B. .044
   C. 22.5
   D. .277
   E. 1.6

9. How many quarts of soda are in a three-liter bottle?
   A. .317
   B. .106
   C. 9.47
   D. 2.85
   E. 3.18

10. One gallon of milk equals
    A. .95 liters
    B. 3.8 liters
    C. 4.25 liters
    D. 1.8 kg
    E. 4.25 qt

For #11–15, answer:

A  if the quantity in column A is greater.
B  if the quantity in column B is greater.
C  if the two quantities are equal.
D  if the relationship cannot be determined from the information given.

Write your answer in the blank.

|     | A     | B        |     |
| --- | ----- | -------- | --- |
| 11. | 2 qt  | 2 liters | ___ |
| 12. | 5 mi  | 5 km     | ___ |
| 13. | X lb  | X kg     | ___ |
| 14. | 2 oz  | 56 g     | ___ |
| 15. | X cm  | Y in     | ___ |

# LESSON TEST 31

Circle your answer.

1. A fractional exponent indicates a:
   A. fraction
   B. radical
   C. square
   D. factor
   E. division problem

2. When multiplying two identical numbers with exponents:
   A. add the exponents
   B. subtract the exponents
   C. divide the exponents
   D. multiply the exponents
   E. raise the first exponent by the power of the second

3. $27^{\frac{1}{3}} =$
   A. 27
   B. 9
   C. 3
   D. 81
   E. 7

4. $(X^3)^{\frac{1}{3}} =$
   A. 9X
   B. $X^2$
   C. $X^6$
   D. $X^9$
   E. X

5. $125^{\frac{2}{3}} =$
   A. 83.3
   B. 25
   C. 15
   D. 5,208
   E. 5

6. $X^{\frac{2}{3}} =$
   A. $(\sqrt[3]{X})^2$
   B. $(\sqrt{X})^3$
   C. $(3X)^2$
   D. $(2X)^3$
   E. $\frac{2X}{3}$

7. $2 \cdot 16^{\frac{1}{4}} =$
   A. $2^{\frac{3}{4}}$
   B. $2^{16}$
   C. 2
   D. 4
   E. 8

8. $\left(Y^{\frac{1}{6}}\right)\left(Y^{\frac{2}{5}}\right) =$
   A. $Y^{\frac{2}{30}}$
   B. $Y^{\frac{17}{30}}$
   C. $30Y^2$
   D. $(Y^2)^{\frac{2}{30}}$
   E. $\frac{2Y}{30}$

ALGEBRA 1 LESSON TEST 31

LESSON TEST 31

9. $\left(x^2 \cdot x^4\right)^{\frac{1}{2}} =$
   A. $x^8$
   B. $x^4$
   C. $x^6$
   D. $x^3$
   E. $x^2$

10. $10^{\frac{2}{3}} \cdot 1{,}000 =$
    A. $10^{\frac{8}{3}}$
    B. $1{,}000$
    C. $10^{\frac{11}{3}}$
    D. $3 \cdot 10^4$
    E. $10^2$

For #11–15, answer

    A if the quantity in column A is greater.
    B if the quantity in column B is greater.
    C if the two quantities are equal.
    D if the relationship cannot be determined from the information given.

(All variables are greater than one.)

Write your answer in the blank.

| | A | B |
|---|---|---|
| 11. | $2^2$ | $\left(2^{\frac{1}{2}}\right)^2$ |
| 12. | $x^3 x^3 x^{\frac{1}{3}}$ | $x^{19}$ |
| 13. | $\left(3^{\frac{1}{3}}\right)^3$ | $\left|\sqrt{9}\right|$ |
| 14. | $10^2 \cdot 1000$ | $10^5$ |
| 15. | $\left[\left(B^3 B^5\right)^{\frac{1}{2}}\right]^2$ | $B^4$ |

# LESSON TEST 32

Circle your answer.

1. Scientific notation is used:
   A. to simplify computations with very large or small numbers
   B. to make work look more scientific
   C. only in astronomy
   D. to practice using exponents
   E. as part of the metric system

2. 6,300,000 in scientific notation is:
   A. $.63 \times 10^7$
   B. $6.3 \times 10^6$
   C. $6.3 \times 10^5$
   D. $6.3 \times 10^{-6}$
   E. $63 \times 10^5$

3. 543,000 in scientific notation is:
   A. $543 \times 10^3$
   B. $54.3 \times 10^{-4}$
   C. $54.3 \times 10^4$
   D. $5.43 \times 10^5$
   E. $5.43 \times 10^{-5}$

4. .00065 in scientific notation is:
   A. $6.5 \times 10^4$
   B. $6.5 \times 10^{-4}$
   C. $6.5 \times 10^{-1}$
   D. $65 \times 10^{-5}$
   E. $6.5 \times 10^{-2}$

5. .0000781 in scientific notation is:
   A. $78.1 \times 10^{-6}$
   B. $781 \times 10^{-7}$
   C. $7.81 \times 10^{-2}$
   D. $7.81 \times 10^5$
   E. $7.81 \times 10^{-5}$

6. $(10^7)(10^7)$ equals:
   A. $100^4$
   B. $100^{10}$
   C. $10^{14}$
   D. $10^{10}$
   E. $10^{21}$

7. $(10^1)(10^{-6})$ equals:
   A. $10^{-6}$
   B. $10^5$
   C. $10^{-5}$
   D. $100^{-6}$
   E. $100^{-5}$

8. 12,000 x .006 can be expressed as:
   A. $(1.2 \times 6)(10^4 \times 10^{-3})$
   B. $(1.2 \times 6)(10^{-4} \times 10^3)$
   C. $(1.2 \times 6)(10^{-12})$
   D. $72(10^4 \times 10^{-3})$
   E. 72,000

ALGEBRA 1 LESSON TEST 32

LESSON TEST 32

9. $3.6 \times 10^{-6}$ can be written as:
   A. 360,000
   B. .000036
   C. .0000036
   D. -360,000
   E. $(3.6^{-6})(10)$

10. $1.02 \times 10^5$ can be written as:
    A. 1,020,000
    B. $5.10 \times 10$
    C. 10,200,000
    D. .0000102
    E. 102,000

11. Multiply $.25 \times 130,000$ using scientific notation.
    A. $3.25 \times 10^5$
    B. $3.25 \times 10^{-5}$
    C. $3.25 \times 10^4$
    D. $32.5 \times 10^4$
    E. $3.25 \times 10^{-6}$

12. Multiply $50,000,000 \times .610$ using scientific notation.
    A. $30.5 \times 10^6$
    B. $3.05 \times 10^{-6}$
    C. $3.05 \times 10^6$
    D. $3.05 \times 10^7$
    E. $.31 \times 10^7$

13. Multiply: $2.4 \times 3.06$. Your final answer should include proper significant digits.
    A. 7.4
    B. 7.34
    C. 7.3
    D. 7.344
    E. 73.4

14. Multiply: $(1.24 \times 10^6) \div (4.7 \times 10^3)$. Your final answer should include proper significant digits and scientific notation.
    A. $5.83 \times 10^9$
    B. $5.828 \times 10^9$
    C. $58.28 \times 10^8$
    D. $5.8 \times 10^9$
    E. $58.3 \times 10^8$

15. Divide: $(6.25 \times 10^8) \div (3.241 \times 10^4)$. Your final answer should include proper significant digits and scientific notation.
    A. $19.28 \times 10^3$
    B. $1.93 \times 10^4$
    C. $1.928 \times 10^4$
    D. $2.0 \times 10^4$
    E. $1.9284 \times 10^4$

# LESSON TEST 33

Circle your answer.

1. In our decimal system, every place value is based on:
    - A. 2
    - B. 5
    - C. 10
    - D. 12
    - E. 100

2. In base 12, the letter "A" stands for:
    - A. 2
    - B. 3
    - C. 10
    - D. 12
    - E. 20

3. When changing a number in another base to base 10, the first step is to:
    - A. divide the number by its base
    - B. multiply the number by its base
    - C. write the number in exponential notation
    - D. multiply by 10
    - E. divide by 10

4. "Four in base 10" is written as:
    - A. $4/10$
    - B. $4^{10}$
    - C. $10^4$
    - D. $10_4$
    - E. $4_{10}$

5. The number four can be written in base four as:
    - A. $10/4$
    - B. $10^4$
    - C. $4^{10}$
    - D. $10_4$
    - E. $4_{10}$

6. To change $300_{10}$ to base 5, first divide by:
    - A. 5
    - B. 25
    - C. 125
    - D. 10
    - E. 100

7. To change $95_{10}$ to base 4, first divide by:
    - A. 64
    - B. 4
    - C. 16
    - D. 10
    - E. 40

8. Change $34_{10}$ to base 4.
    - A. $402_4$
    - B. $220_4$
    - C. $202_4$
    - D. $22_4$
    - E. $420_4$

ALGEBRA 1 LESSON TEST 33

LESSON TEST 33

9. Change $45_{10}$ to base 2.
   A. $21011_2$
   B. $10011_2$
   C. $10101_2$
   D. $1011_2$
   E. $101101_2$

10. Change $356_{10}$ to base 12.
    A. $251_{12}$
    B. $258_{12}$
    C. $7101_{12}$
    D. $215_{12}$
    E. $25A_{12}$

11. Change $122_6$ to base 10.
    A. $60_{10}$
    B. $732_{10}$
    C. $18_{10}$
    D. $48_{10}$
    E. $50_{10}$

12. Change $4B3_{12}$ to base 10.
    A. $576_{10}$
    B. $711_{10}$
    C. $744_{10}$
    D. $579_{10}$
    E. $43_{10}$

13. Change $52A_{12}$ to base 10.
    A. $754_{10}$
    B. $96_{10}$
    C. $64_{10}$
    D. $888_{10}$
    E. $624_{10}$

14. $122_3$ is the same as:
    A. $244_6$
    B. $122_3$
    C. $36_{10}$
    D. $38_{10}$
    E. $17_{10}$

15. $1000_7$ is the same as:
    A. $1000^7$
    B. $343_{10}$
    C. $34{,}000_{10}$
    D. $21_{10}$
    E. $7{,}000_{10}$

# LESSON TEST 34

Circle your answer.

1. $X^2 + Y^2 = R^2$ is the equation of a(n):
   A. circle
   B. line
   C. ellipse
   D. square
   E. none of the above

2. $AX^2 + BY^2 = C$ is the equation of a(n): ($A \neq B$)
   A. circle
   B. line
   C. ellipse
   D. square
   E. none of the above

3. In the equation $(X - h)^2 + (Y - k)^2 = R^2$, h and k represent:
   A. the measurements of an ellipse
   B. the coordinates of the center of a circle
   C. points on a line
   D. a point on the circumference of a circle
   E. none of the above

4. If $X^2 + Y^2 = P^2$, then:
   A. P is the diameter of a circle.
   B. $P^2$ is the radius of a circle.
   C. P is the radius of a circle.
   D. P is the Y-intercept of a line.
   E. none of the above

5. How can the standard equation of an ellipse be distinguished from the standard equation of a circle?
   A. Only the circle has different X and Y coefficients.
   B. The ellipse has only the X term squared.
   C. Only the circle has any squared terms.
   D. Only the ellipse has squared terms.
   E. Only the ellipse has different X and Y coefficients.

6. Given $X^2 + Y^2 = 9$, what is the center of the circle?
   A. (3, 3)
   B. (1, 9)
   C. (0, 0)
   D. (X, Y)
   E. (9, 9)

7. Given $X^2 + Y^2 = 9$, what is the radius of the circle?
   A. 3
   B. 9
   C. 1
   D. 5
   E. 81

8. Given $(X - 3)^2 + (Y - 4)^2 = 16$, what is the center of the circle?
   A. (−4, −4)
   B. (4, 4)
   C. (−3, −4)
   D. (4, 3)
   E. (3, 4)

ALGEBRA 1 LESSON TEST 34

LESSON TEST 34

Use this diagram for #9 –15.

9. Which figure represents
   $(X + 3)^2 + (Y + 3)^2 = 4$?
   A. Q
   B. R
   C. S
   D. T
   E. U

10. Which figure represents
    $(X - 3)^2 + (Y - 2)^2 = 1$?
    A. Q
    B. R
    C. S
    D. T
    E. none

11. Which figure represents
    $(X + 3)^2 + (Y - 3)^2 = 1$?
    A. P
    B. Q
    C. R
    D. T
    E. none

12. Which figure represents
    $(X - 3)^2 + (Y + 3)^2 = 4$?
    A. P
    B. Q
    C. T
    D. U
    E. none

13. Which figure represents
    $X^2 + 4Y^2 = 4$?
    A. P
    B. S
    C. R
    D. U
    E. none

14. Which figure represents
    $4X^2 + Y^2 = 4$?
    A. P
    B. S
    C. R
    D. U
    E. none

15. Which figure represents
    $X^2 + Y^2 = 1$?
    A. Q
    B. R
    C. S
    D. T
    E. none

ALGEBRA 1

# LESSON TEST 35

Circle your answer.

1. $3X^2 + 4Y^2 = 12$ is the equation of a(n):
   A. line
   B. circle
   C. ellipse
   D. parabola
   E. hyperbola

2. $XY = 16$ is the equation of a(n):
   A. line
   B. circle
   C. ellipse
   D. parabola
   E. hyperbola

3. $X^2 + Y^2 = 36$ is the equation of a(n):
   A. line
   B. circle
   C. ellipse
   D. parabola
   E. hyperbola

4. $2Y + 3X + 5 = 0$ is the equation of a(n):
   A. line
   B. circle
   C. ellipse
   D. parabola
   E. hyperbola

5. $Y = 4X^2$ is the equation of a(n):
   A. line
   B. circle
   C. ellipse
   D. parabola
   E. hyperbola

6. In which quadrants will the graph for $XY = 8$ appear?
   A. I and II
   B. I and III
   C. II and IV
   D. I only
   E. III and IV

7. In which quadrants will the graph for $XY = -8$ appear?
   A. I and II
   B. II and III
   C. II and IV
   D. I only
   E. III and IV

8. In $Y = AX^2$, if the value of A increases, which is true?
   A. The radius of the circle gets larger.
   B. The parabola gets wider.
   C. The parabola shifts upward on the Y-axis.
   D. The parabola gets narrower.
   E. The parabola gets farther from the origin.

9. Given $XY = 12$, if X is $-4$, what is the value of Y?
   A. $-3$
   B. 3
   C. 8
   D. $-8$
   E. $-12$

ALGEBRA 1 LESSON TEST 35

LESSON TEST 35

Use these diagrams for #10–15.

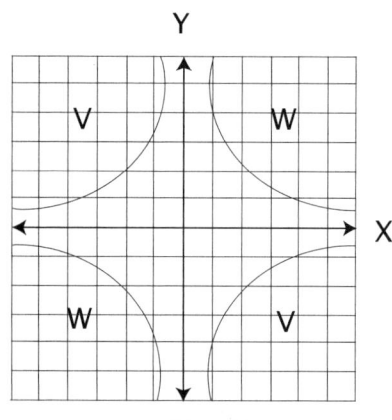

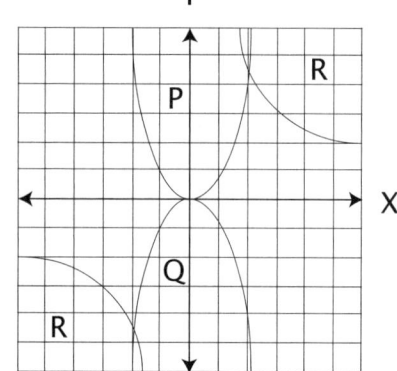

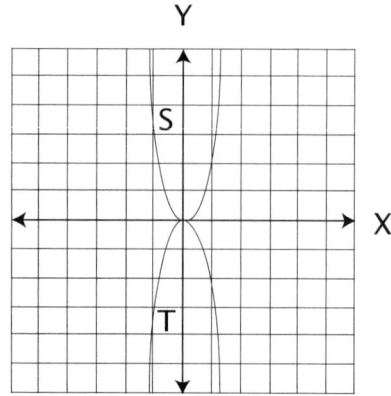

10. Which figure represents
 XY = 4?
 A. P
 B. R
 C. S
 D. V
 E. W

11. Which figure represents
 XY = –4?
 A. P
 B. R
 C. T
 D. V
 E. W

12. Which figure represents
 XY = 9?
 A. Q
 B. R
 C. S
 D. V
 E. W

13. Which figure represents
 $Y = X^2$?
 A. P
 B. Q
 C. R
 D. S
 E. T

14. Which figure represents
 $Y = 3X^2$?
 A. P
 B. R
 C. S
 D. T
 E. W

15. Which figure represents
 $Y = -X^2$?
 A. P
 B. Q
 C. R
 D. S
 E. T

UNIT TEST Lessons 24-35 (100 points possible)

I. Divide and check. (8 points each)

1. $X+2 \overline{\smash{)}2X^2+5X+2}$

2. $X-2 \overline{\smash{)}X^2+3X^2-9X-2}$

II. Find the factors. (5 points each)

1. $3X^2 - 12$

2. $Q^2 - R^2$

3. $2X^2 - 4X - 30$

III. Solve by factoring, and then check. (6 points each)

1. $X^2 + 5X + 16 = 10$

2. $2X^3 - 18X = 0$

ALGEBRA 1 UNIT TEST III

UNIT TEST III

IV. Using the table, convert the following measures. (4 points each)

> 1 mi ≈ 1.6 km    1 km ≈ .62 mi
> 1 oz ≈ 28 g      1 g ≈ .035 oz

1. Change 100 ounces to grams.

2. Change six kilometers to miles.

V. Write each number using scientific notation. (3 points each)

1. 456,700,000

2. .0260

VI. Solve using scientific notation. (4 points each)

1. .0003 x 4.2

2. $\dfrac{6,800,00}{200,000}$

UNIT TEST III

VII. Simplify. (3 points each)

1. $\sqrt{196}$

2. $\sqrt{100A^2}$

3. $\sqrt{X^2+18X+81}$

VIII. Change to the base indicated. (4 points each)

1. $70_{10}$ = _____ $_7$

2. $2210_3$ = _____ $_{10}$

IX. Simplify. (3 points each)

1. $16^{\frac{1}{2}}$

2. $(1{,}000)^{\frac{2}{3}}$

ALGEBRA 1          91

UNIT TEST III

X. Graph the following and identify the graph as a circle, ellipse, parabola, or hyperbola. You may add more points to the chart if you wish. (6 points each)

1. $XY = -1$

| X | Y |
|---|---|
| 1 | |
| -1 | |
| $\frac{1}{2}$ | |
| $-\frac{1}{2}$ | |
| $\frac{1}{3}$ | |
| $-\frac{1}{3}$ | |

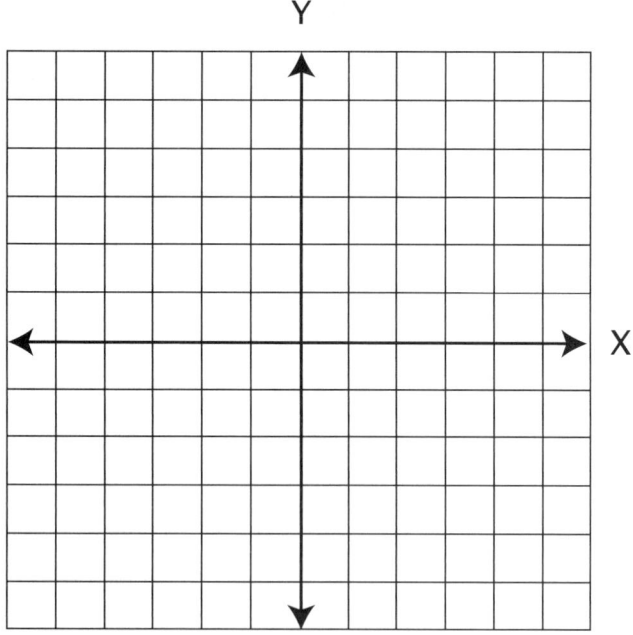

2. $X^2 + Y^2 = 4$

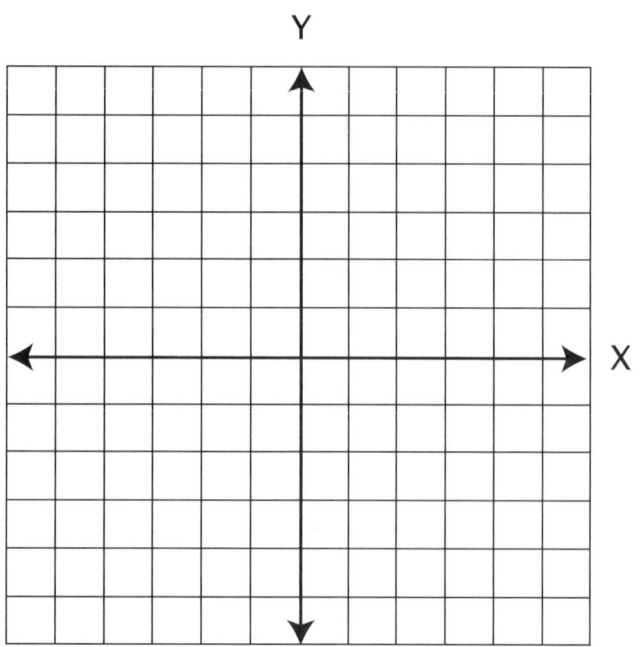

FINAL EXAM (150 points possible)

I. Express in simplest terms. (6 points each)

1. $\left(-\dfrac{1}{2}\right)^2 + (ab)^0 - 3^2$

2. $\sqrt{16x^2}$

3. $\left(2^2\right)^3 \left(2^2\right)$

4. $|6 - 8|$

5. $\sqrt{x^2 + 4x + 4}$

6. $(81)^{\frac{1}{2}}$

7. $\dfrac{3x^2}{x^{-4}} + \dfrac{5x}{x^{-1}} =$

ALGEBRA 1 FINAL EXAM

FINAL EXAM

II. Factor. (10 points each)

1. $3X^2 - 27$

2. $5X^2 - 9X - 2$

3. $X^3 + 5X^2 + 6X$

4. $14Y^2 - 7Y - 42$

III. Solve for X. (8 points)

1. $10^6 = (10^3)^X$

IV. Solve for X. Factor first if necessary. (10 points each)

1. $3X^2 - 6X = 0$

2. $\dfrac{1}{6}X - \dfrac{1}{2} = \dfrac{2}{3}$

3. $X + .25X = .4$

V. Graph. (10 points each)

1. $Y = 2X^2$

| X | Y |
|---|---|
| 0 | |
| 1 | |
| -1 | |
| 2 | |
| -2 | |

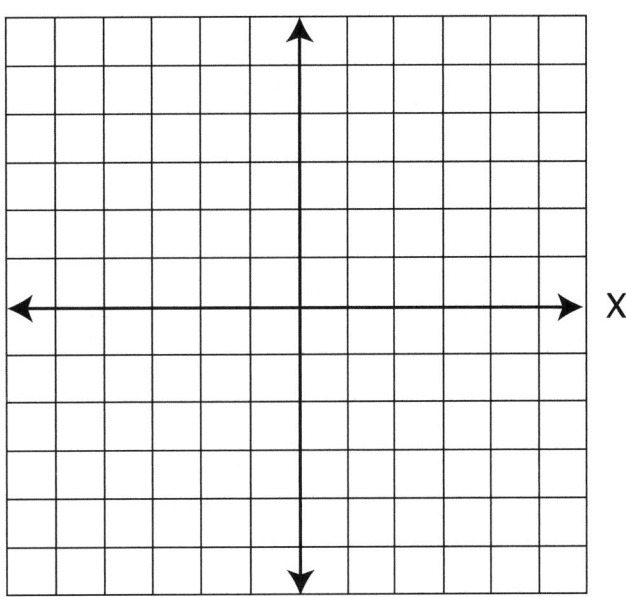

2. $4X^2 + Y^2 = 16$

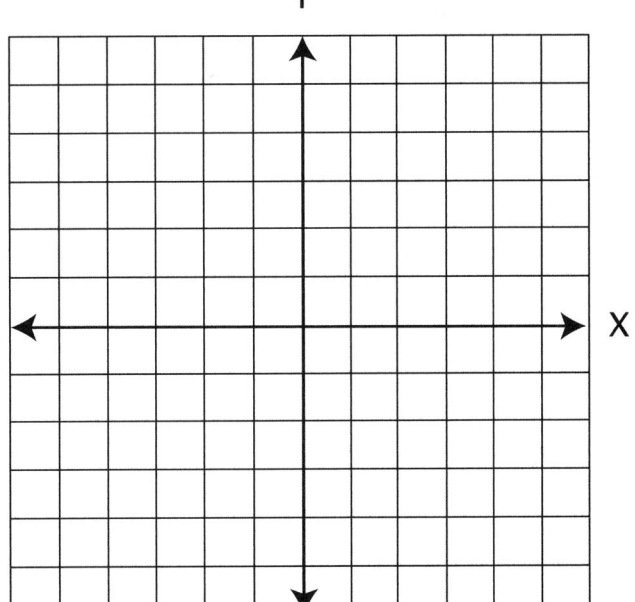

FINAL EXAM

3.  Y = 3X + 1

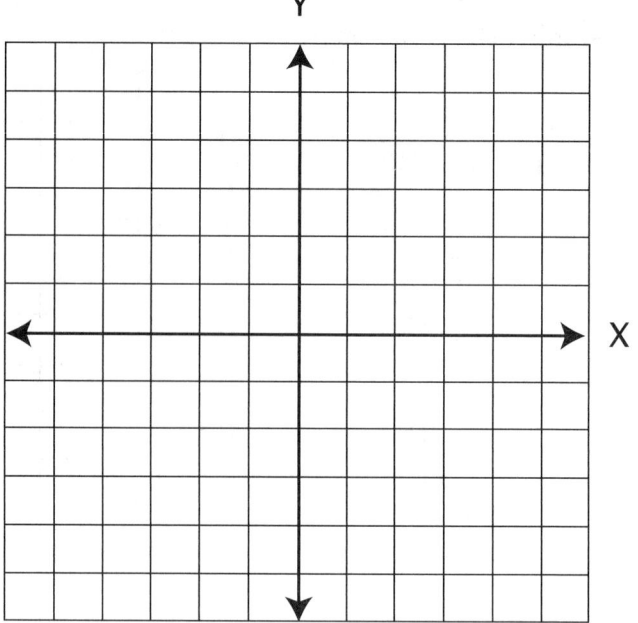